PARAMETER OPTIMIZATION IN FRICTION SURFACING

Author

(S.GODWIN BARNABAS)

Preface to the book

Surface engineering deals with the surface of the solid matter and it is sub-discipline of materials science. Solids are composed of a bulk material covered by a surface and it is called surface phase. The surface phase of a solid interacts with the surrounding environment. This interaction can degrade the surface phase over time, may result in loss of material from its surface. Environmental degradation of the surface phase over time can be caused by wear, corrosion, creep, fatigue loads, shear loads, tensile loads, cutting forces or when exposed to higher temperature. Major types of wear include abrasion, friction, erosion and corrosion. Wear can be minimized by modifying the surface properties of solids by surface finishing or by use of lubricants.

Author

(S.GODWIN BARNABAS)

LIST OF CONTENTS

ABOUT THE AUTHOR

S.GODWIN BARNABAS is working as a Assistant professor in Velammal College of Engineering and Technology, Madurai. He has published 13 international journals, 10 international conferences and he has authored 5 books for LAMBERT PUBLICATIONS .His field of interest is Recycling Management

CHAPTER 1

INTRODUCTION

1.1 Surface Engineering

Surface engineering deals with the surface of the solid matter and it is sub-discipline of materials science. Solids are composed of a bulk material covered by a surface and it is called surface phase. The surface phase of a solid interacts with the surrounding environment. This interaction can degrade the surface phase over time, may result in loss of material from its surface. Environmental degradation of the surface phase over time can be caused by wear, corrosion, creep, fatigue loads, shear loads, tensile loads, cutting forces or when exposed to higher temperature. Major types of wear include abrasion, friction, erosion and corrosion. Wear can be minimized by modifying the surface properties of solids by surface finishing or by use of lubricants.

The corrosion will affect not only the metals, but also the non-metals like plastics, rubber, ceramics etc. Corrosion effects the safe, reliable and efficient operation of equipment or structures. Corrosion is a slow process and except noble metals such as gold and platinum, there is no material which will withstand corrosive attack in all environments. Corrosion affects both the performance and the life of devices which are used in various fields.

Surface engineering involves altering the properties of the solid surfaces which could be different from those of the core material to reduce the degradation over time. These are also used to impart a wide range of functional properties, including physical, chemical, electrical, electronic, magnetic, mechanical, wear-resistant and corrosion-resistant properties at the required substrate surfaces. Almost all types of materials, including metals, ceramics, polymers, and composites can be coated on materials, similar or dissimilar.

This is accomplished by making the surface robust to the environment in which it will be used to protect from wear and corrosion to prolong its life and also enhance aesthetic appearance of the substrates.

Surface engineering techniques are being used in the automotive, aerospace, missile, power, electronic, biomedical, textile, petroleum, petrochemical, chemical, and steel, power, cement, machine tools, and construction industries.

1.2 Friction Surfacing

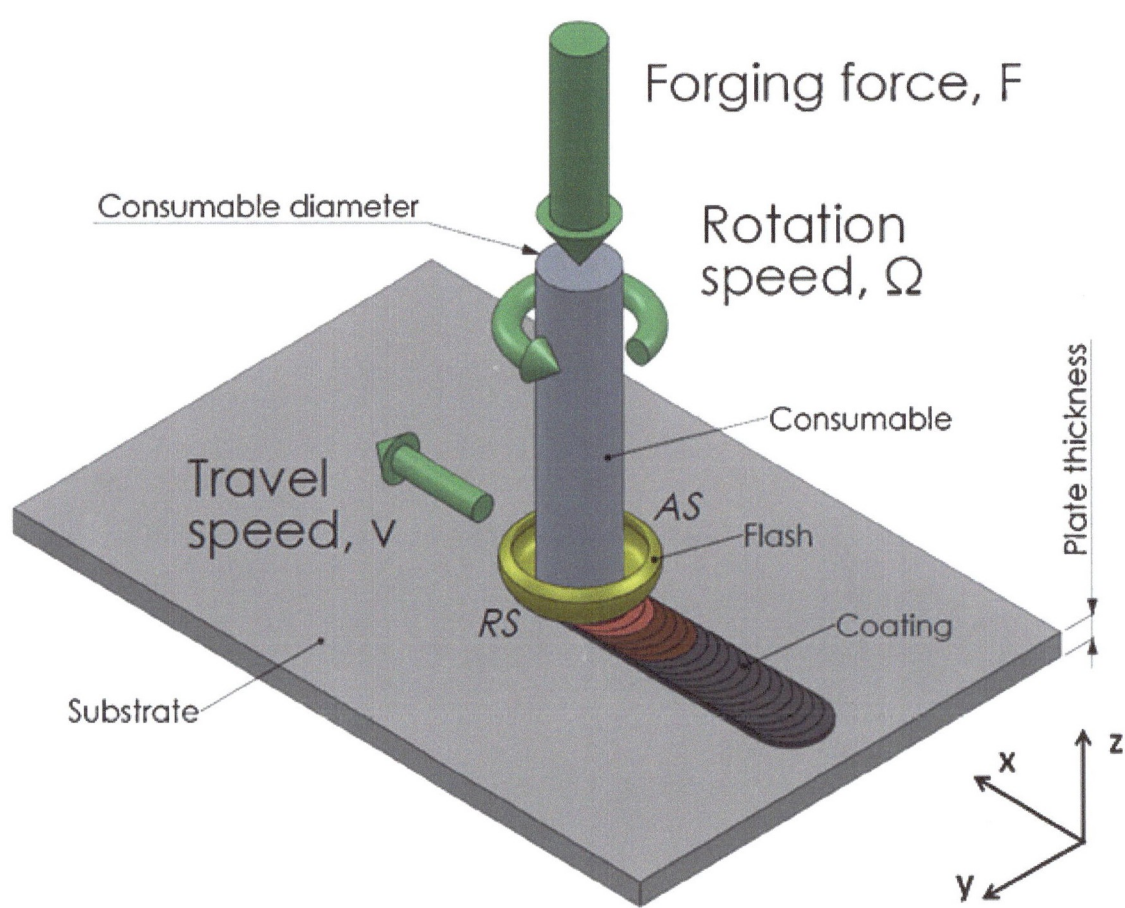

Fig 1.1 Schematic of friction surfacing

Surface engineering has become a relevant research field for manufacturing industries, as it enables advanced component design and a selective functionalization of surfaces. Solid state processing technologies are now mature and reliable alternatives to conventional processes, as stated by Mishra and Ma (2005).

Friction surfacing is a promising new technology for depositing metallurgically bonded coatings on engineering components to combat wear and corrosion. Being a solid state process, friction surfacing elimates the problems such as porosity, hot cracking, segregation, and dilution which are commonly associated with fusion-based techniques. This is attained because no melting is involved in this process.

Hard facing /coating techniques based on fusion welding and thermal spraying are generally employed to protect steel surface from corrosion. But fusion welding based coating techniques generally suffers from dilution and thermal spraying results in mechanical bonding rather than metallurgical bonding

Research so far has revealed that in friction surfacing the mechatrode force (F), mechatrode rotation speed (N) and substrate traverse speed (V_x) are of critical importance for the final quality of the coating and bond.

In the present study, three state variables that reflect coating quality were considered as a subject for optimisation and in this context a target for process parameter selection. These are coating thickness (C_t), coating width (C_w). The optimisation procedure considered in this study involved.

➤ Development of a methodology for in-process precision measurement of axial load, traverse speed and rotational speed.

➤ Development of an empirical model involving process parameters of coating quality state variables i.e coating thickness and coating width.

The friction surfacing machine consists of a power rotor which can move vertically with high precision Z. Under the rotor there is an XY table, which can be positioned and moved accurately. The system is controlled using a serial computer link. The input parameters to the machine being:

- Spindle rotation speed
- Spindle direction
- Table movement

1.2.1 Principle of Friction Surfacing

In this process, rotating cylindrical consumable rod is fed against a substrate with axial force acting simultaneously on the rod. The frictional heat is generated between the substrate and the consumable rod. Once the rubbing end of the consumable rod is sufficiently plasticized, the substrate is traversed horizontally with respect to the vertical consumable rod. Material flow at the area of contact occurs due to the combined effect of axial load, rod rotational speed, and substrate traverse speed. As the substrate moves at a specific rate, the plasticized metal deposits over it. The vertical force consolidates the plasticized metal and results in the formation of a continous and metallurgically bonded layer. The width and thickness of the track thus produced on the substrate depend on mechatrode material and diameter as well as friction surfacing process parameters (axial force, mechatrode rotational speed and substrate traverse speed) and hence are called critical process parameters. Usually, the width is 0.9 times as the diameter of the mechatrode.

The material properties such as thermal conductivity, sliding characteristics, plasticity and physical properties such as mechatrode diameters will influence the thickness of deposit. The presence of high contact stress between substrate and mechatrode, removes oxide films over the substrate surface. The process is, to a large extent, self cleaning and regulating.

1.2.2 Mechanism of Bonding and Properties

Friction surfacing process can deposit a wide variety of materials and alloys, with ideal metallurgical bond, onto a range of metal substrates. In friction welding process, the metals are joined with combination of axial force and the friction heat generated between the moving surfaces of metals, shearing and forging mechanism is involved for causing plasticity in the mechatrode. Hence selection of the process parameters depends on the properties of the materials being joined. The process is a solid phase bonding method, there is negligible dilution and the bond has good strength. Since softening of the zone of interface occurs, the deposit will have fine grain microstructure due to rotary forging action. The ductility of the deposit will be higher than the consumable rod.

The tensile strength and elongation will be lowered but equal to the ultimate tensile strength of weaker metal due to severe softening action. This has to be tolerated while using for any application.

1.2.3 Material Combinations of Friction Surfacing

Friction surfacing is used to join a wide range of similar and dissimilar materials such as, stainless steel over mild steel, aluminum and its alloys over steels, high alloy hard facing deposits over steels, inter-metallic, MMC(metal matrix composite), and many of the dissimilar metal combinations that cannot be joined by conventional fusion welding techniques. Friction surfacing is most suitable to join similar and dissimilar metals depending on the design and of economic considerations. Hence it is an alternate method used where other processes are not preferable due to widely differing properties of dissimilar metals.

Friction surfacing with titanium, aluminum and brass on mild steel pose difficulties. Both brass and aluminum consumables fail to form a heated layer in contact with the mild steel due to high thermal conductivity of either metal.

1.2.4 Advantages of Friction Surfacing

Process is reliable and repeatable

Excellent bonding with no inclusions, porosity or oxidation

Coating material properties improved

Dense, clean & fine microstructures

'Non-weldable' (super) alloys can be deposited

No melting of materials

Negligible dilution

- Small localized HAZ (heat affected zone)
- No cracking in the HAZ
- Similar or dissimilar materials can be bonded

1.2.5 Applications of Friction Surfacing

Several applications have been conceived for this unique process. A wide range of component sizes and shapes have been surfaced for critical end use. Friction surfacing is the unique solid phase bonding process that facilitates metallurgical bonds, with applications in the areas of wear resistance, protection against corrosion, aesthetic appearance to the surfaces, homogeneous repairs to valuable components and to achieve desired thermal/electrical properties. The development of the friction surfacing also leads to cladding of pressure vessels with titanium, aluminum and stainless steel.

The critical areas of application include depositing hard facing materials on cutting edges of knives of various categories, punch, die, tools and blades required for food processing, chemical, agriculture and medical industries. Friction surfacing is most suitable for reconditioning of worn out shafts, in-situ reclamation of worn out railway points and good anti-corrosion overlay of slide valve plates.

Friction surfacing not only gives good bond on plane surfaces but also on other contours by design of special purpose machines using CNC technology. Since bond strength is very good, these deposits are expected to serve better during service. There is also some advancement in using robots for wear and corrosion resistant deposits on 3D complex geometric surfaces. Friction surfacing is being developed to use under water for marine propulsion system. This process is yet to be commercialized in India.

CHAPTER 2
LITERATURE REVIEW

H.Khalid Rafi, G.D.Janaki Ram, G.Phanikumar and K.Prasad Rao [1] studied the effects of traverse speed on the geometry, interfacial bond characteristics and mechanical properties of coatings .

M.Chandrasekaran, A.W.Batchelor and S.Jana [2] studied that mild steel bonded well with the substrate and there was evidence of interfacial compound formation whereas in case of stainless steel there was no evidence of mixing and coating.

G.Madhusudhan Reddy and T. Mohandas [3] studied that stainless steel coating of mild steel leads to the formation of carbides in the stainless steel adjacent to the interface as a result of carbon migration from mild steel towards stainless steel.

J.John Samuel Dilip and G.D.Janaki Ram [4] studied the individual layers upto the thickness of 1mm to 2mm can be added up successively by friction deposition. A solid cylinder of 20mm diameter and 50mm height was successfully produced with austenitic stainless steel AISI 304.

H. Khalid rafi, N.Kishore babu, G.Phanikumar and K.Prasad Rao [5] studied the microstructural evolution of stainless steel AISI 304 on low carbon steel using optical microscopy,electron back scattered diffraction and transmission electron microscopy.

Ramesh Puli, E. Nandha Kumar and G.D. Janaki Ram [6] showed that the microstructure tests showed good hardness results when stainless steel is coated over mild steel. Bend and shear tests indicated excellent coating/substrate bonding.

J.Gandra, R.M.Miranda and P.Vilac [7] studied the influence of axial force, rotation and traverse speed on interfacial bond properties were investigated.

G.M. Bedford, V.I. Vitanov and I.I. Voutchkov [8] studied the mechanism of auto hardening of the mechatrode coating on substrate is studied.

B.Jaworski, G.M.Bedford, I.Voutchkov and V.I.Vitanov [9] studied the procedures for data collection, management and optimization of friction surfacing process and found that the thickness of the coated layer is typically between 0.5-3mm depending on the mechatrode material and diameter.

V.I.Vitanov, I.I.Voutchkov and G.M.Bedford [10] studied the three state variables, that reflect coating quality were considered as a subject for optimization and in this context a target for process parameter selection which are coating thickness, coating width, coating bond strength.

M.Chandrasekaran, A.W.Batchelor and S.Jana [11] studied that a nominal contact pressure as high as 21.9Mpa was required to obtain an adherent coating of uniform quality for mild steel with tool steel and inconel.

D.Govardhan, A.C.S.Kumar, K.G.K.Murti and G.Madhusudhan Reddy [12] studied the effect of process parameters such as frictional pressure, rotational speed of the mechatrode and welding speed .Their interaction effects on the deposit for the consumable rod are identified.

M. chandrasekar, A.W.Batchelor and S.Jana [13] found that tool steel and inconel were efficiently deposited onto steel to form a dense strong coating while aluminium was deposited only at high contact pressures. Titanium could not be deposited under the tested conditions.

K Prasad Rao, A.Veera Sreenu and Krishnan Balasubramaniam [14] carried out a thermography study on tool steel and copper coatings by friction surfacing. Infrared thermography is used to record and analyze the thermal profile.

X.M. Liu, Z.D. Zou, Y.H. Zhang, S.Y. Qu and X.H. wang [15] studies revealed that the material at the top of the coating rod is plastic and that it behaves as a quasi-liquid in the friction surfacing process.

Ramesh Puli and G.D. Janaki Ram [16] found that friction surfaced coatings contains a fully martensitic micro-structure showing comparable wear and corrosion resistance.

H. Khalid rafi, G.D.Janaki Ram, G.Phanikumar and K.Prasad Rao [17] studies revealed that coating exhibited martensitic micro structure with fine grain size and with no carbide particles. Coatings in as deposited conditions showed a very high hardness compared to the mechatrode material in annealed condition.

H. Khalid rafi, G.D.Janaki Ram, G.Phanikumar and K.Prasad Rao [18] results showed that coating width is a strong function of mechatrode rotational speed while coating thickness is mainly dependant on substrate traverse speed. Lower mechatrode rotational speed results in wider coating and higher substrate traverse speed produce thinner coatings .Thinner coating exhibits higher bond strength than thicker coatings.

Rick Greenough [19] found that friction surfacing process is also suitable for depositing hard and wear resistant material, cladding welded joints in corrosion resistant pressure vessels such as titanium clad, aluminum clad and stainless steel clad. The maximum temperature reached in friction surfacing is reached is lower than the other surfacing methods such as weld overlay (Electric arc welding, plasma arc spraying, flame spraying and laser methods).

Hoshihiro Yamashita and Kazuhiro [20] found that friction surfacing process is most useful to do in situ repairs of aged structural material in nuclear plants (LWR Reactors. The maximum temperature attained in friction surfacing is found to be 1200° C which is lower than the melting point of stainless steel.

CHAPTER 3
GENERAL PROPERTIES OF MATERIALS USED

3.1 Stainless Steel AISI 304

It is the most versatile and the most widely used of all stainless steels. Its chemical composition, mechanical properties, weld ability and corrosion/oxidation resistance provide the best all-round performance stainless steel at relatively low cost.

Stainless steel is 100% recyclable. 50% of new stainless steel is made from re-melted scrap metal, rendering it an eco-friendly material.

3.1.1 Chemical Composition of Stainless Steel AISI 304

Table 3.1

Material	C	Ni	Mn	S	P	Cr	Fe
%Composition	0.08	8-10.5	2	0.03	0.045	18-20	Remainder

3.1.2 Mechanical Properties of Stainless Steel AISI 304 at Elevated Temperature

Table 3.2

Temperature (°c)	UTS (MPa)	YS (MPa)	% of elongation
Room	586/615*	241/318*	55/62*
204	496	159	51
316	469	134	45
427	441	114	40
538	386	97	36
649	303	88	34
760	200	76	40
871	110	-	-

3.1.3 Physical Properties of Stainless Steel AISI 304

Density: 8.03 g/cm^3, Specific Heat: 0.50 (kJ/ kg• K), Melting Range: 1399 – 1454°C and Thermal Conductivity at 100°C: 16.2(W/m •K) and 21.4(W/m •K) at 500°C.

3.1.4 Applications of Stainless Steel AISI 304

- Process equipment in the mining, chemical, cryogenic, food, dairy and   pharmaceutical industries.

- Tanks and containers for storing large variety of liquids and solids.

- In the marine environment, it is also used for nuts, bolts, screws, and other fasteners.

3.2 Spheroidal Graphite Iron

S.G Cast iron is defined as a high carbon containing, iron based alloy in which the graphite is present in compact, spherical shapes rather than in the shape of flakes, the latter being typical of gray cast iron . As nodular or spheroidal graphite cast iron, sometimes referred to as ductile iron, constitutes a family of cast irons in which the graphite is present in a nodular or spheroidal form. The graphite nodules are small and constitute only small areas of weakness in a steel-like matrix. Because of this the mechanical properties of ductile irons related directly to the strength and ductility of the matrix present—as is the case of steels.

While most varieties of cast iron are brittle, ductile iron has much more impact and fatigue resistance, due to its nodular graphite inclusions.

3.2.1 Metallurgical characteristics of Spheroidal Graphite Iron

Ductile iron is not a single material but is part of a group of materials which can be produced to have a wide range of properties through control of the microstructure. The common defining characteristic of this group of materials is the shape (morphology) of the graphite. In ductile irons, the graphite is in the form of nodules rather than flakes (as in grey iron), thus inhibiting the creation of cracks and providing the enhanced ductility that gives the alloy its name. The formation of nodules is achieved by addition of nodulizing elements, most commonly Magnesium (note Magnesium boils at 1100C and Iron melts at 1500C) and, less often now, Cerium (usually in the form of Misch metal). Tellurium has also been used. Yttrium, often a component of Misch metal, has also been studied as a possible nodulizer.

3.2.2 Chemical Composition of Spheroidal Graphite Iron

Table 3.3

Material	C	Si	Mn	S	Mg	Fe
% composition	3-4	1.8-2.8	0.1-1	0.03 max	0.01-0.10	remainder

3.2.3 Physical Properties of Spheroidal Graphite Iron

Density: 7.1 g/cm^3, Specific Heat: 0.51 (kJ/ kg• K), and Thermal Conductivity of ferritic grades (20-500 °C) is 36 W/m °K. Conductivity for pearlitic grades over the same temperature range is approximately 20 per cent less.

3.2.4 Applications of Spheroidal Graphite Iron

- Ductile iron in the form of ductile iron pipe is used for water and sewer lines.

- Support bracket for agricultural tractor.

- Tractor life arm.

- Check beam for lifting track.

- Mine cage guide brackets.

- Gear wheel and pinion blanks and brake drum.

- Machines worm steel.

- Flywheel.

- Thrust bearing.

- Frame for high speed diesel engine.

- Four throw crankshaft.

- Fully machined piston for large marine diesel engine.

- Bevel wheel.

- Hydraulic clutch on diesel engine for heavy vehicle.

- Fittings overhead electric transmission lines.

- Boiler mountings, etc.

3.2.5 Advantages of Spheroidal Graphite Iron over Mild Steel

Whether specifying a new casting, forging or fabrication or improving an existing one, the potential benefits of Ductile Iron over mild Steel are clear.

- Improved strength to weight ratio.

- Better surface definition and finish.

- Reduced machining allowance.

- Reduced Component Cost.

CHAPTER 4
EXPERIMENTAL WORK AND GEOMETRY MEASUREMENT

 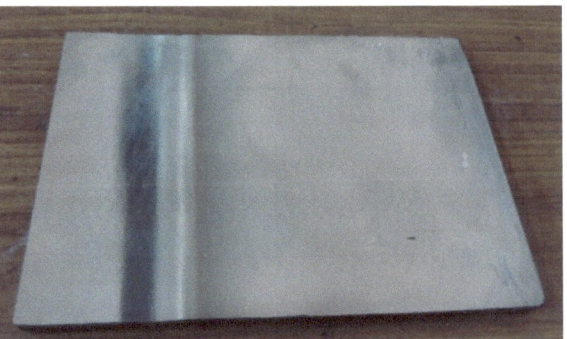

Fig 4.1 Substrate before grinding   **Fig 4.2 Substrate after grinding**

4.1 Specimen Preparation (Substrate)

Step 1: Making to a proper dimension

Initially the ductile iron plate was of 1500*1000*8 mm dimension. After cutting the raw material with gas arc cutter, it became 100*150*8 mm which is perfect for our friction surfacing process.

Step 2: Rough finishing by emery paper

Initially the ductile iron material got from shop is fully corroded. But with corroded surface friction surfacing will not be good, and thus must be removed. Emery is a type of paper that can be used for sanding down hard and rough surfaces. Even after hard rubbing with emery paper the ductile iron plate is still corroded. Hence fine finishing with surface grinding machine is a must after rough finishing.

Step 3: Fine finishing with surface grinding

This is used to get a fine finish over the roughly finished surface obtained by emery paper. Surface grinding machine is being used.

Step 4: Applying acetone solution over the surface.

The surface of the SG iron is cleaned with acetone. Acetone removes all impurities like oil, grease, dust etc.

Step 5: Corrosion free surface

Finally we get a ductile iron plate without any corrosive layer. After the fine finis0hing process followed by acetone cleaning, our material is completely ready to use for friction surfacing.

4.2 Specimen Preparation (Mechatrode)

- 22mm diameter 304 stainless steel rod is cut into 105 mm length pieces.
- These rods are turned by holding between the centers of lathe to get uniform 20mm in diameter with a 100 mm length.

4.3 Parameters Involved

- The load given on mechatrode in kN is directly proportional to the bonding efficiency. In higher loads the external energy required to form a coating is less.
- The rotating speed given on mechatrode in RPM is directly proportional to the time taken for plasticization of metals. In higher rotating speeds plasticization occurs easily.
- The traverse speed given on base metal in mm/sec is inversely proportional to the coating thickness. In lower traverse speed heat affected zone will be more.

4.4 Factors Affecting Selection of Process Parameters

- The maximum friction pressure depends on forging strength of the consumable rod and the substrate materials used.
- Coefficient of friction at the interface of consumable rod and substrate from room temperature to melting point of consumable.
- The physical properties like thermal conductivity and specific heat also influence the levels of parameters.
- The temperature at the interface of stainless steel and ductile iron should be nearer to the melting point of the stainless steel. But quantum of heat generated depends on frictional pressure, rotational speed and torque-time characteristics.
- Generally the width of the deposit is 0.8 times the diameter of the mechatrode and gives good integrity and adhesion strength and it is in the range of 60-90% of the mechatrode diameter.
- This process is suitable for the consumables which are having less thermal conductivity than substrate and poor sliding characteristics.
- Dwell time for initial rubbing is of 5 seconds and is selected for this material combination.

4.5 Formation of Orthogonal Array by Taguchi Method

No. of parameters: 3   No. of levels: 3

Selection of array: L9 (3 * 3) orthogonal array

Here 'L' signifies the No. of levels and '9' signifies the No. of runs. Thus L9 array has 9 runs. A L9 array can be used to cover all combination of two parameters at three levels each. The general structure of the L9 array is as shown below: **Table 4.1**

Trial. No	A	B	C
T1	1	1	1
T2	1	2	2
T3	1	3	3
T4	2	1	2
T5	2	2	3
T6	2	3	1
T7	3	1	3
T8	3	2	1
T9	3	3	2

The total no. of runs in the Taguchi method is nine with three replicates.

The L9 array structure is substituted with the parametric values of the project at the respective cells to form the L9 array of the project for optimization.

Table 4.2

Trial. No	Axial Load (kN)	Rotational Speed (rpm)	Traverse Speed (mm/s)
T1	7	800	1.6
T2	7	1200	2.2
T3	7	1600	2.7
T4	8	800	2.2
T5	8	1200	2.7
T6	8	1600	1.6
T7	9	800	2.7
T8	9	1200	1.6
T9	9	1600	2.2

Result and Discussion

Fig 4.3 Experimental setup of friction surfacing machine

4.6 Experimental Work with Friction Surfacing Machine

Mechatrode stainless steel is fitted in a mechatrode holder, which consists of splines at its outside surface along with its axis to allow the movement in the axial direction while doing friction surfacing. The mechatrode holder is fixed in a spindle and locked by using threading. In this condition the mechatrode can move along with the axis and simultaneously rotate with the spindle. Bush is fitted at the end of the spindle with locking screws to allow the mechatrode to rotate the axis of
 the spindle perfectly, under axial load.

The process parameters as per treatment combination are set on the computer of the machine. Dwell time of 5 seconds is found to get the mechatrode reach plastic state as per the initial trials conducted before start of experimental trials.

Machine is started and, once the consumable is sufficiently heated to acquire forging temperature, the welding speed is automatically switched on. The hot consumable material flows plastically over the substrate to form a coating. Since the machine is designed to deposit the consumable material in one direction of table, after completion of required length of the weld, the consumable which is fitted in the spindle automatically detach from the substrate by moving spindle in upward direction by stopping immediately the spindle rotation and welding speed.

4.7 Friction Surfaced Sample Photos

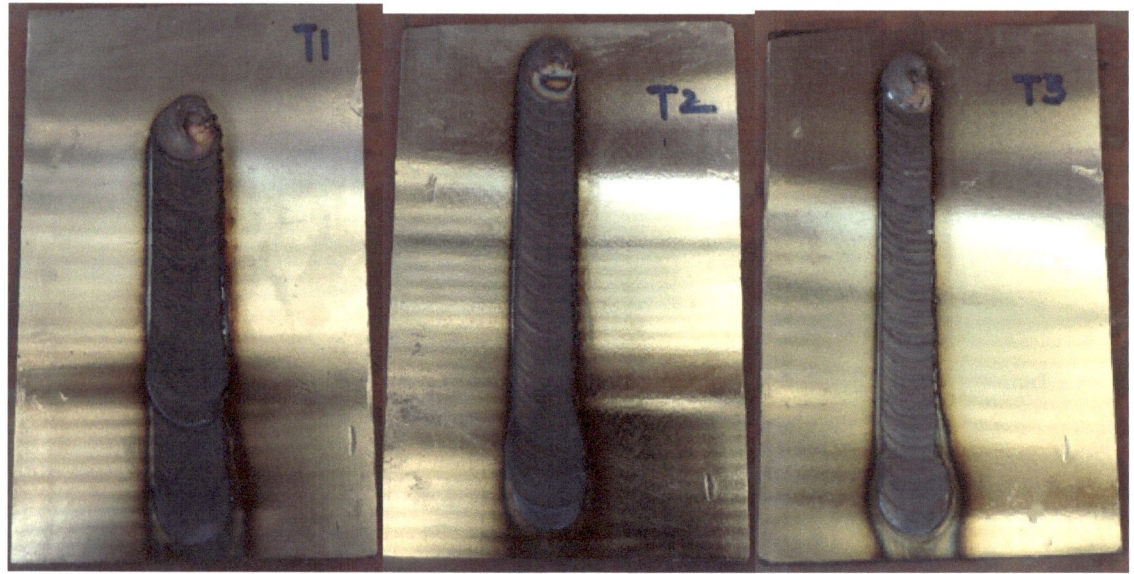

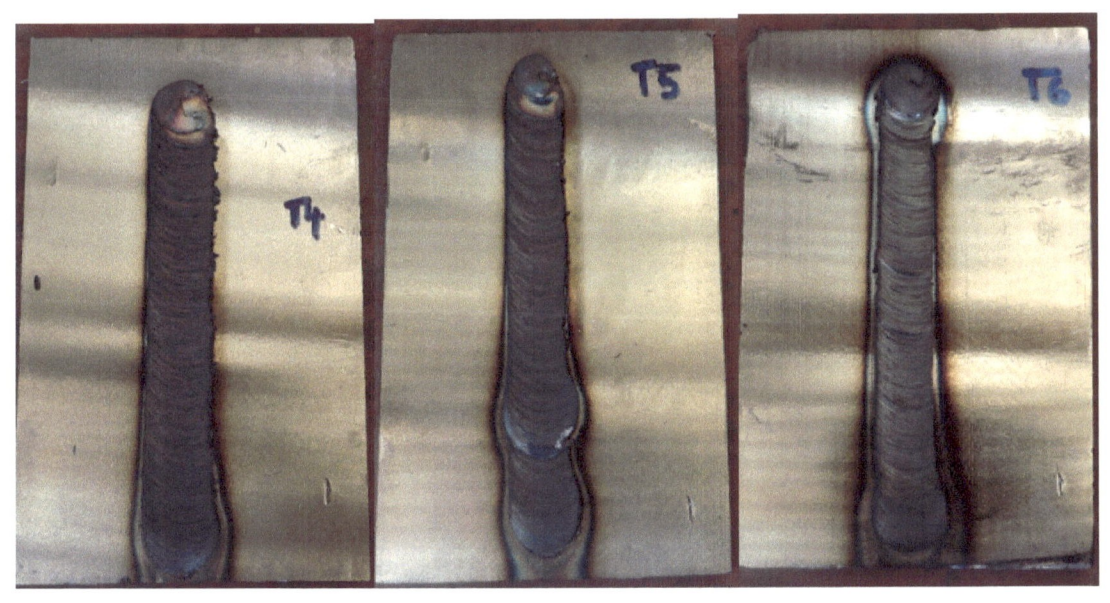

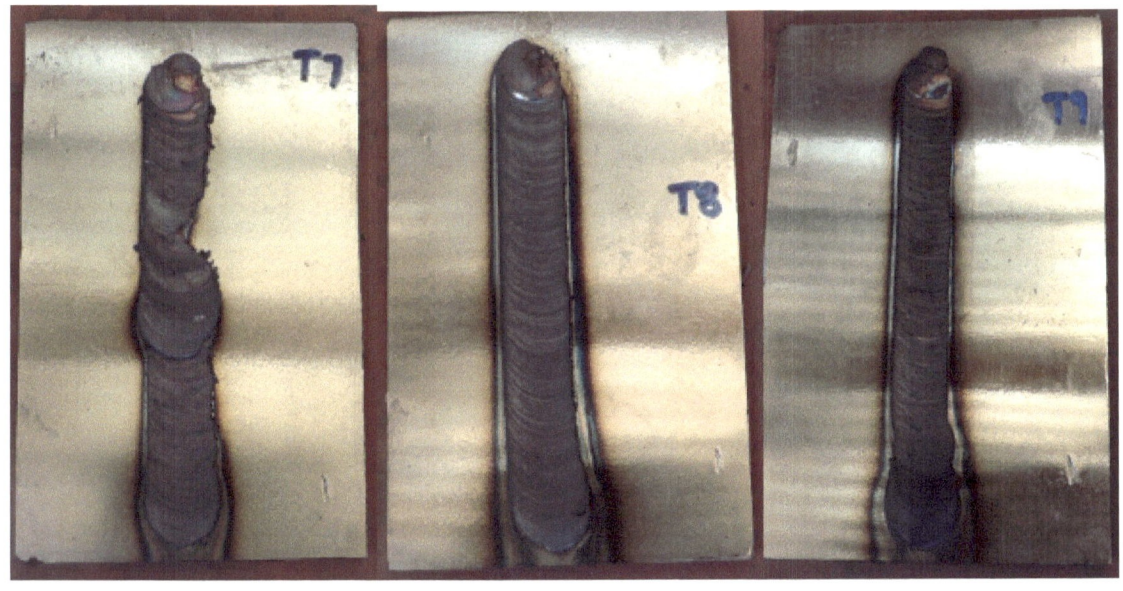

4.7.1 Coating Thickness (C_t) on Deposit Geometry

Table 4.3

22

S. No	Substrate	Mechatrode	Coating Thickness c_t (mm)
1			2.9
2			2.36
3			1.45
4			2.73
5	Sg Iron	Stainless Steel 304	2.21
6			2.36
7			2.33
8			2.55
9			1.95

- In stainless steels, the coating thickness is inversely proportional to the traverse speed.
- In higher traverse speed, the time of deposition of plasticized material on the work piece is less. Hence gives less coating thickness and less heat affected zone in work piece.
- Coating with minimum thickness is more advisable automobile parts applications. Higher coating thickness will give increase in weight of the component.

4.7.2 Coating Width (C_W) on Deposit Geometry

Table 4.4

S. No	Substrate	Mechatrode	Coating Width c_t (mm)
1			19.72
2			16.83
3			14.49
4			18.9
5	Sg Iron	Stainless Steel 304	17.29
6			16.14
7			19.23
8			19.44
9			16.02

- The width of the flash formed in the substrate is usually 0.9 times the diameter of the mechatrode used.
- Our results show approximately the same value.

4.7.3 Length of Mechatrode

Fig 4.4 Length of mechatrode after friction surfacing

Table 4.5

Mechatrode	Set No	Length	
		Before(mm)	After(mm)
Stainless Steel 304	1	100	69
	2	100	67
	3	100	79
	4	100	72
	5	100	71
	6	100	65
	7	100	70
	8	100	62
	9	100	69

- According to the machine specification the length of the mechatrode rod must be in 90-120mm.

- Difference in mechatrode length before and after experiment shows the material consumption during the process.
- The loss of volume of material during the process is equal to the volume of coating.

4.7.4 Diameter of Mechatrode

Fig 4.5 Diameter of mechatrode after friction surfacing

Table 4.6

Mechatrode	Set No	Diameter	
		Before(mm)	After(mm)
Stainless Steel 304	1	20	38
	2	20	33
	3	20	35
	4	20	34
	5	20	33
	6	20	35
	7	20	35
	8	20	36
	9	20	34

- According to the machine specification the diameter of the mechatrode rod must be in 16-24mm.
- During the process material temperature reaches above 700^0C
- In that high temperature plasticization occurs. Some material melts and deposited on the work piece called coating.
- The remaining material stick around the mechatrode edge, after some time in the atmospheric air get cooled and looks bigger in diameter than the initial diameter.

Chapter 5

INSPECTION AND TESTING RESULTS

5.1 Hardness Measurement

Hardness of the obtained samples is tested using a micro Vickers hardness tester. The specifications of which are as follows

Machine Name	:	Micro Vickers Hardness Tester
Testing load range	:	10 grams to 1 Kg Load
Make	:	Wilson Wolpert – Germany
Micrometer least count	:	0.01 mm
Hardness testing Scales	:	HV, HR"A", HR"B", HR"C", 15N, 30N & 45N, 15T, 30T & 45T

Hardness Values in H.V. @ 0.5 Kg load.

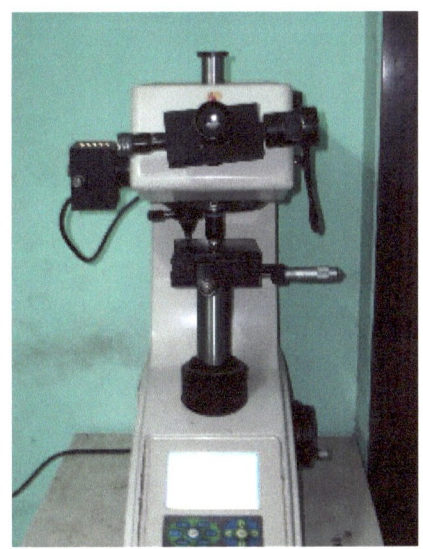

Fig 5.1 Micro hardness test apparatus

Hardness measurements were carried out in the samples sectioned in transverse direction.

Diamond intender is used in this machine. Tests were carried at a load of 0.5 Kg and the following results were obtained.

Table shows the hardness results

Trial 1

Table 5.1

From the Weld	Weld Ss side	S.g Iron Side
Edge	197.5	155.9
0.1	225.4	152.0
0.2	223.3	141.8
0.3	220.9	149.8
0.4	225.0	147.1
0.5	223.0	149.1
0.6	218.2	149.1
0.7	214.8	149.6
0.8	213.2	148.2
0.9	210.3	148.2
1.0	200.2	151.2
1.1	197.5	164.3
1.2	207.6	150.4
1.3	208.1	149.6
1.4	208.1	151.3

Trial 2

Table 5.2

From the Weld	Weld Ss side	S.g Iron Side
Edge	201.6	153.4
0.1	227.2	151.8
0.2	227.3	148.1
0.3	224.3	148.8
0.4	225.9	147.6

0.5	224.5	146.7
0.6	225.7	147.8
0.7	219.8	146.1
0.8	215.9	147.7
0.9	215.5	147.4
1.0	212.4	148.7
1.1	215.9	147.1
1.2	212.8	150.9

Trial 3

Table 5.3

From the Weld	Weld Ss side	S.g Iron Side
Edge	200.4	158.4
0.1	225.8	153.2
0.2	226.6	152.2
0.3	227.4	151.9
0.4	225.7	152.2
0.5	225.1	149.5
0.6	222.7	148.5
0.7	219.4	147.2
0.8	218.8	147.0

0.9	216.3	148.8
1.0	214.8	147.8
1.1	112.6	146.0
1.2	203.5	156.8

Trial 4

Table 5.4

From the Weld	Weld Ss side	S.g Iron Side
Edge	270.7	174.9
0.1	271.9	168.3
0.2	249.2	143.8
0.3	236.5	159.6
0.4	239.9	163.3
0.5	229.7	158.8
0.6	231.1	154.1
0.7	229.4	1598
0.8	225.8	166.4
0.9	221.1	153.7
1.0	226.3	159.4
1.1	203.8	154.2
1.2	201.7	149.6

Trial 5

Table 5.5

From the Weld	Weld Ss side	S.g Iron Side
Edge	271.4	170.4
0.1	271.4	160.2
0.2	244.6	141.2
0.3	239.8	151.9
0.4	235.3	161.2
0.5	224.5	156.5
0.6	222.0	156.5
0.7	221.8	155.2
0.8	220.9	160.0
0.9	216.5	154.8
1.0	204.4	154.8
1.1	199.9	157.0
1.2	200.8	156.8

Trial 6

Table 5.6

From the Weld	Weld Ss side	S.g Iron Side
Edge	275.1	172.4
0.1	273.3	165.2
0.2	247.8	140.2
0.3	232.5	151.9
0.4	239.7	149.2
0.5	224.1	156.5
0.6	219.5	141.5
0.7	229.0	151.2

0.8	212.5	150.0
0.9	221.3	158.8
1.0	238.9	151.8
1.1	197.3	155.0
1.2	216.6	152.8

Trial 7

Table 5.7

From the Weld	Weld Ss side	S.g Iron Side
Edge	231.4	136.3
0.1	134.7	153.4
0.2	232.7	154.8
0.3	229.4	155.4
0.4	217.7	157.6
0.5	216.3	153.5
0.6	206.7	152.8
0.7	209.5	155.4
0.8	204.3	151.7
0.9	202.5	155.2
1.0	200.3	153.3
1.1	198.2	158.2

Trial 8

Table 5.8

From the Weld	Weld Ss side	S.g Iron Side
Edge	235.3	135.3
0.1	138.6	156.8
0.2	234.3	157.1
0.3	229.6	155.4
0.4	217.4	154.6
0.5	202.8	153.3
0.6	205.4	151.8
0.7	207.8	151.3
0.8	200.2	152.6
0.9	201.5	152.2
1.0	194.3	155.4
1.1	196.2	156.3

Trial 9

Table 5.9

From the Weld	Weld Ss side	S.g Iron Side
Edge	233.0	138.1
0.1	133.1	152.9
0.2	233.2	153.3
0.3	229.5	154.8
0.4	214.3	153.7
0.5	206.9	153.0
0.6	204.5	151.3
0.7	207.8	150.8
0.8	201.9	151.0

0.9	200.8	152.4
1.0	197.3	154.1
1.1	198.0	156.5

- Hardness is the property of a material that enables it to resist plastic deformation, usually by penetration.

- Hardness value of material is directly proportional to the strength of that material

5.2 Bend Test

Mechanical testing machine is used for this purpose. The point to be noted is that the test is carried out as per ASTM E-8. Here the sample is cut to the required dimensions so that it can be held. The sample is placed at the top of the support. Following which uniform load is applied at its centre. The sample starts bending. The point where it is about to break is noted and the corresponding load is the maximum that it can withstand. The following is a mechanical testing machine.

Fig 5.2 Bend test apparatus

T1

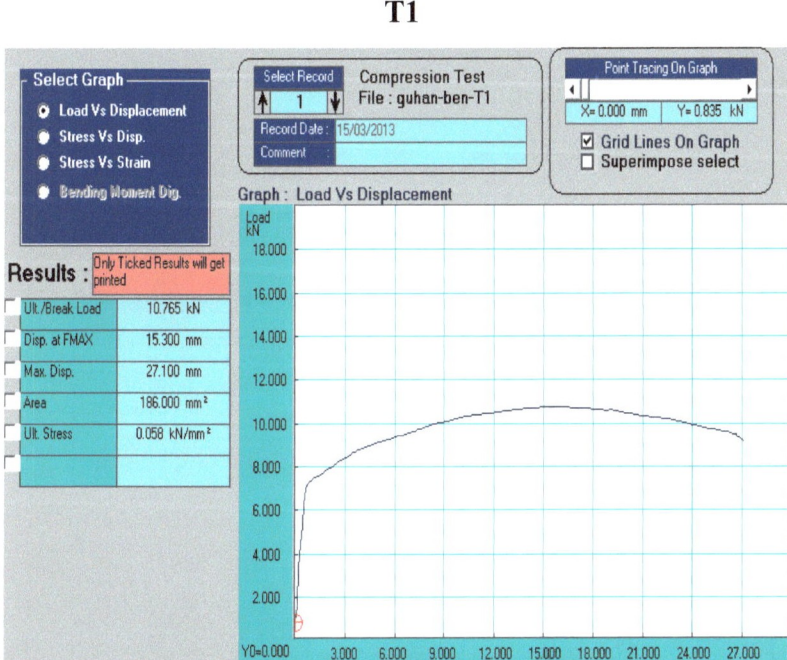

Fig 5.3 Load Vs Displacement Graph for Bend Test

Table shows the bend test results

Table 5.10

S. No	Max Load (kN)	Max Displacement (mm)
1	10.765	27.1
2	10.825	26.9
3	10.729	26.6
4	10.211	23.1
5	10.195	24.3
6	10.301	24.0

7	10.002	27.1
8	10.114	27.4
9	10.065	26.8

- Bend testing determines the ductility or the strength of a material.
- Experiment results show that the strength of a work piece has increased after the coating. Because coating surface provide resistance towards bending.

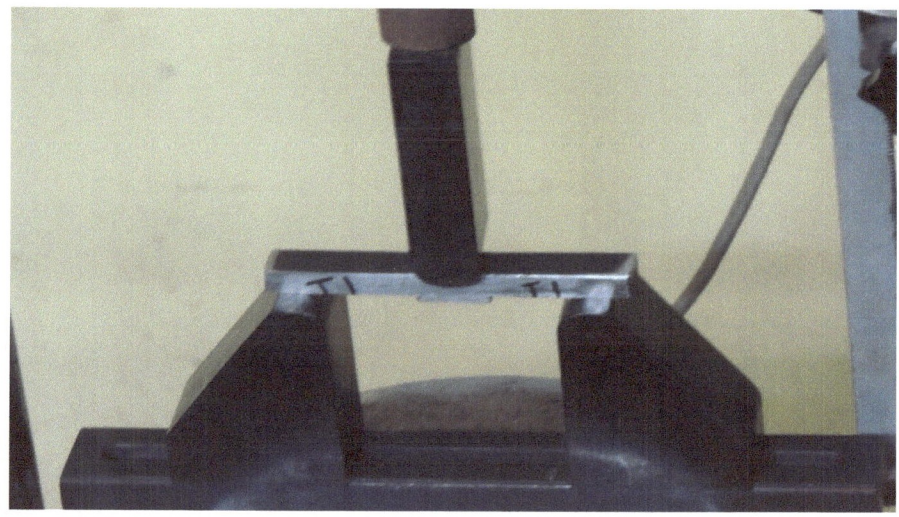

Fig 5.4 Sample before the bend test

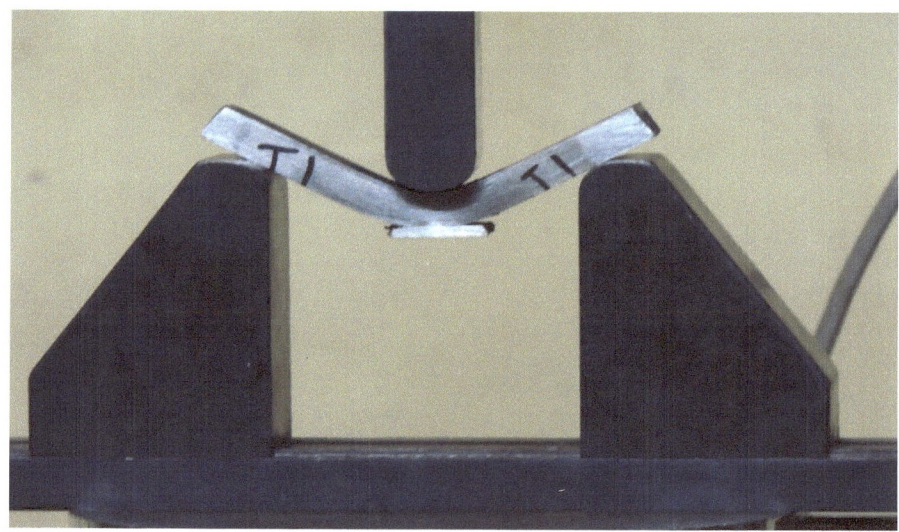

Fig 5.5 Sample after the bend test

- Fig shows the work piece before the load acted on it while under-going the bending test.
- Fig shows the work piece after the load is acted on it.

5.3 Microstructure

Sample T1

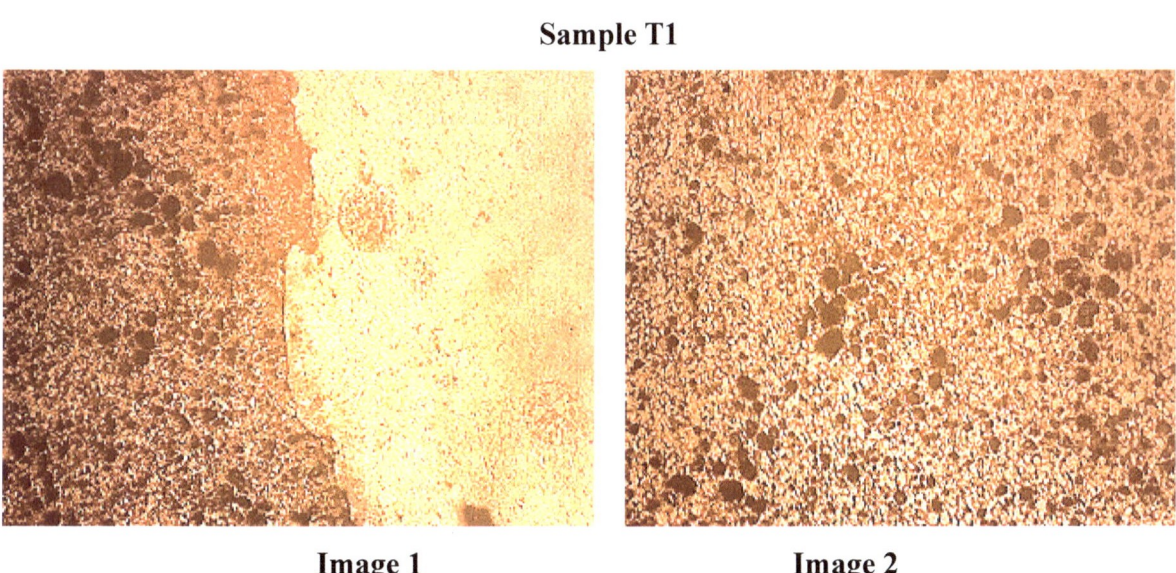

Image 1   **Image 2**

Image 1: shows the base metal SG iron with surfaced SS by friction. The base metal shows fine spheroidal graphite's in ferrite-pearlite matrix.

Image 2: shows the base metal microstructure (SG Iron).

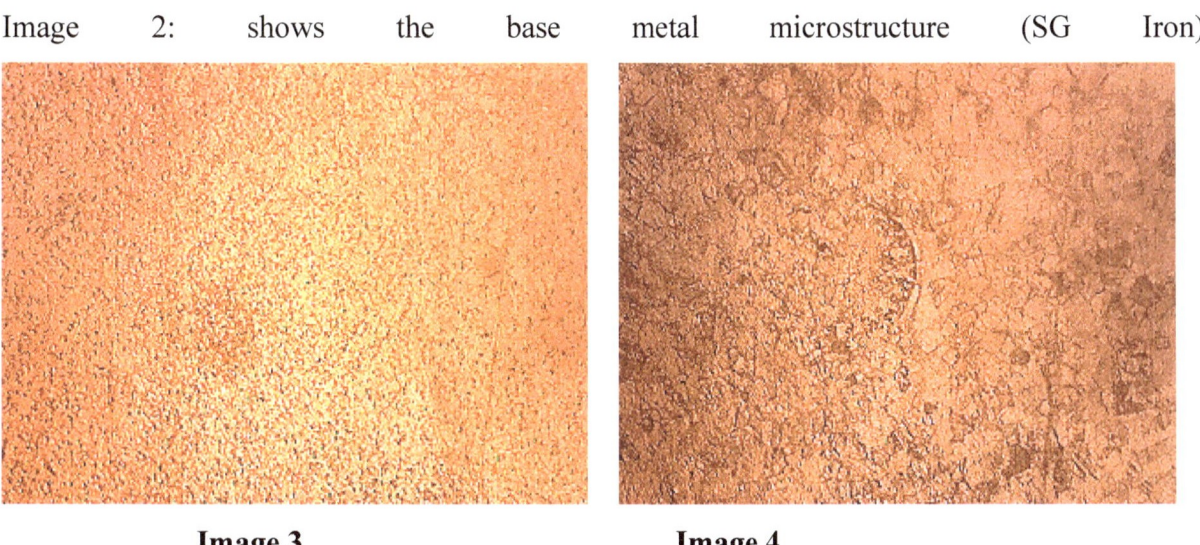

Image 3   **Image 4**

Image 3: shows the etched surfaced SS matrix with fine austenite grains.

Image 4:shows the same SS matrix at higher magnification.

Sample T5

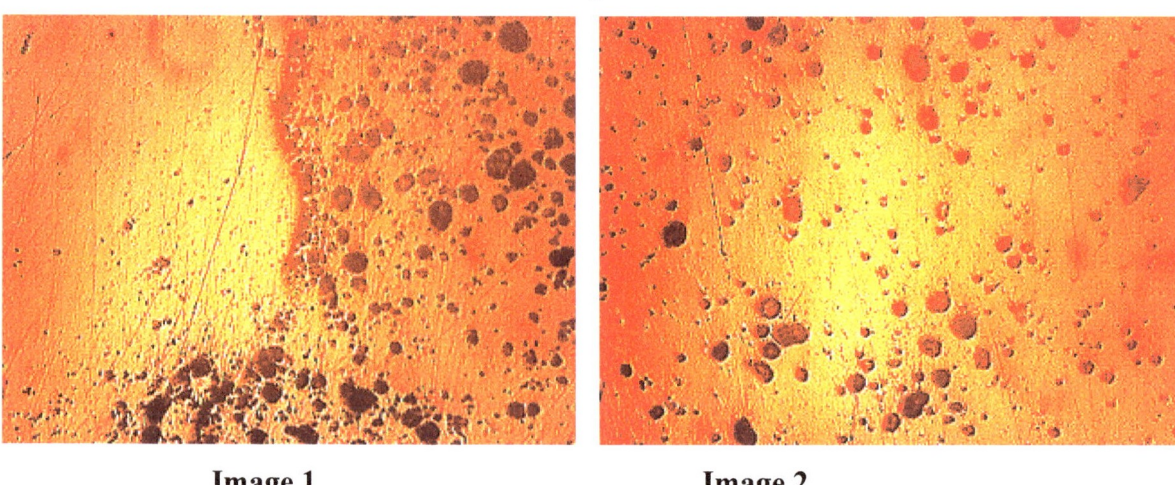

Image 1   **Image 2**

Image 1: shows the base metal SG iron with surfaced SS by friction. The base metal shows fine spheroidal graphite's in ferrite-pearlite matrix.

Image 2: shows the base metal microstructure (SG Iron).

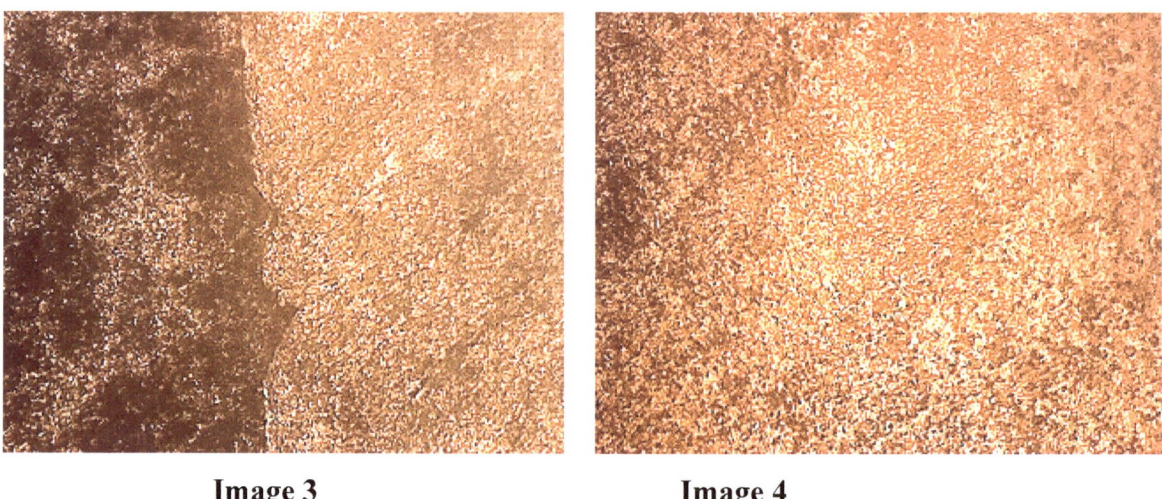

Image 3   **Image 4**

Image 3: shows the base metal SG iron with surfaced SS by friction. The base metal shows fine spheroidal graphite's in ferrite-pearlite matrix. The surfaced metal shows fine austenite grains.

Image 4: shows the same SS matrix at higher magnification.

Sample T9

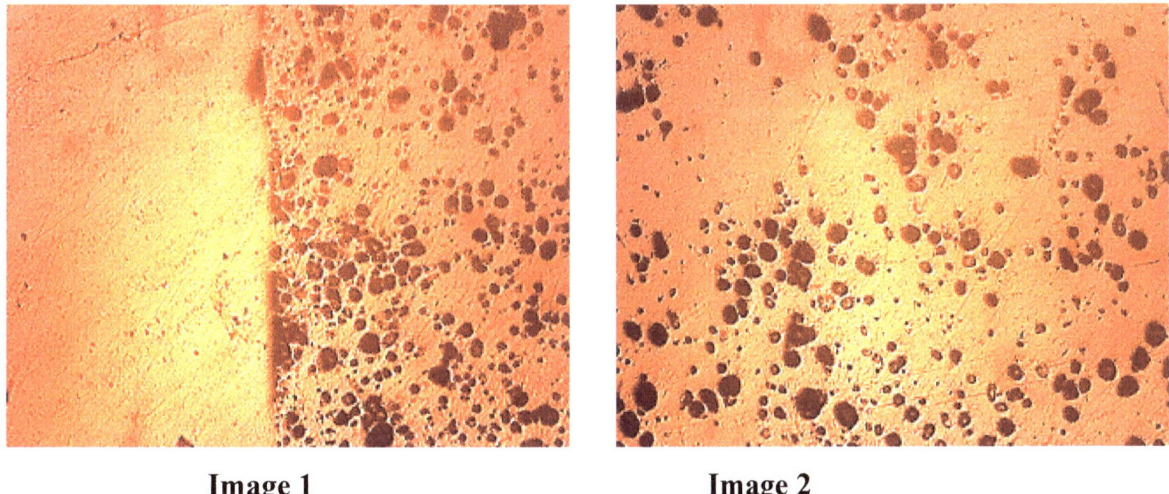

Image 1   **Image 2**

Image 1: shows the base metal SG iron with surfaced SS by friction. The base metal shows fine spheroidal graphite's in ferrite-pearlite matrix. Surfaced metal not etched.

Image 2: shows the base metal microstructure (SG Iron).

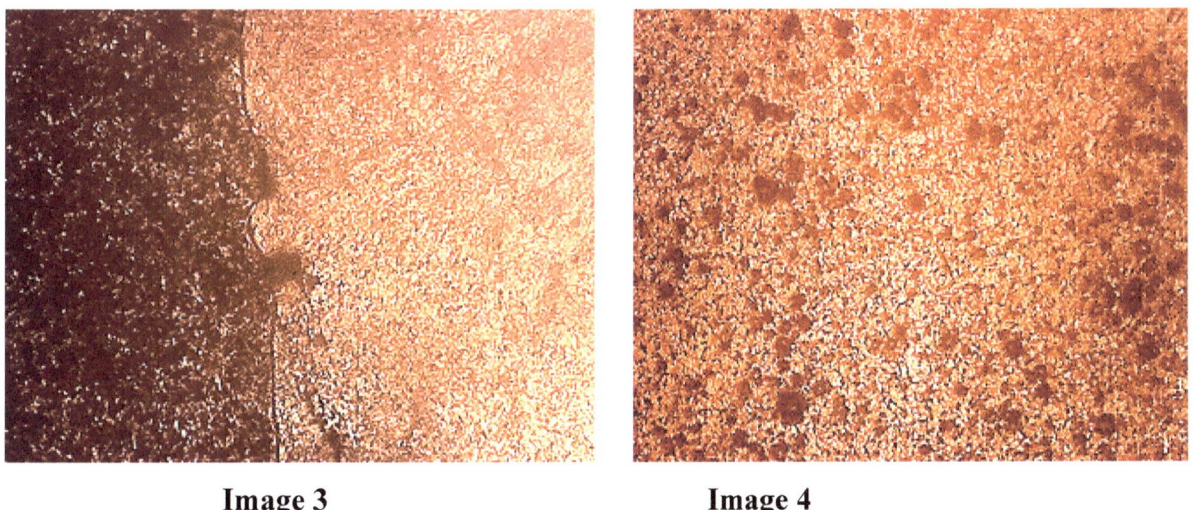

Image 3   **Image 4**

Image 3: shows the base metal SG iron with surfaced SS by friction. The base metal shows fine spheroidal graphite's in ferrite-pearlite matrix. The surfaced metal shows fine austenite grains.

Image 4: shows the base metal microstructure (SG Iron) with ferrite-pearlite matrix.

5.4 Corrosion Test: As per ASTM B117M (Salt Spray fog test)

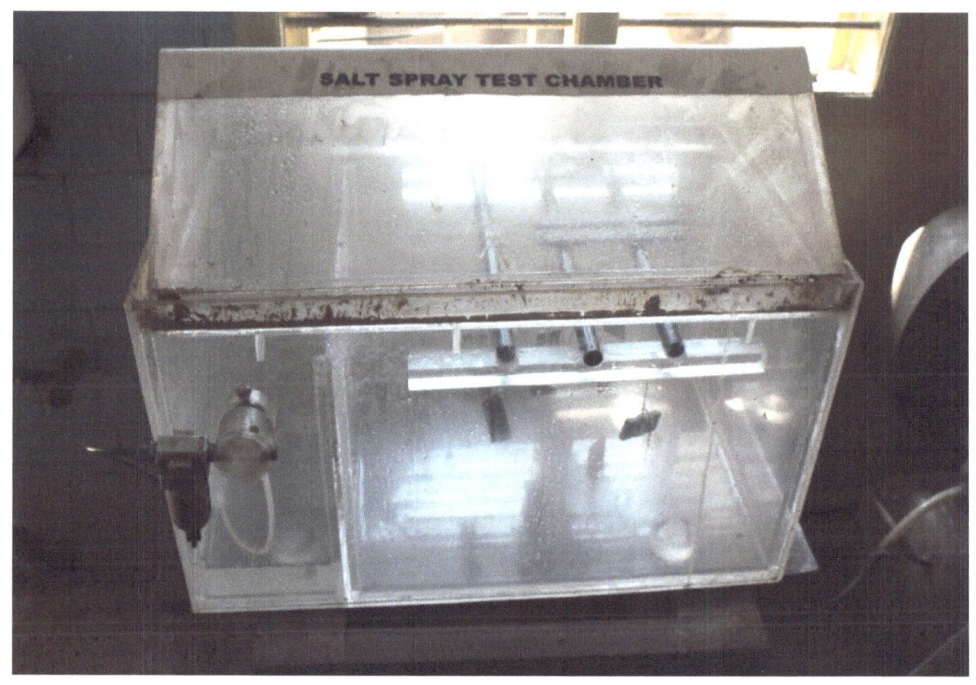

Fig 5.6 Salt spray test chamber apparatus

5.4.1 Salt Spray Apparatus

Specimens are first cut to the size as specified by ASTM. Holes are drilled at the top of each specimen so that they could be held steadily. The weight of the individual specimens are noted before the test starts. The chamber has a provision at the side for spraying NaCl. Sodium chloride is sprayed in the form of fine droplets similar to fog. The purpose of spraying NaCl is because the chlorine atoms react with the coating material of individual specimens and causes its removal. It is sprayed for about 48hours. Following which the specimen is carefully removed and washed with distilled water. Now it is stirred in alcohol on a warm base. After the residues are dissolved it is once again weighed. Comparison of weights before and after the test is now done. The following satisfactory results were obtained.

5.4.2 Salt Spray Test Parameters

1) Temperature of the test: 33 degrees Centigrade.
2) Concentration of the Salt solution: 1.0M
3) Air pressure: 2.0 Kg per Sq. Centimeters.
4) Ph of the Solution followed :   7.0
5) Humidity of the chamber: 95% to 98%.
6) Exposure Time: 48 Hours.
7) Post cleaning: Cleaned in distilled water followed

by rinsing in alcohol

5.4.3 Weight Details for Corrosion Test

Table 5.11

	T1	T5	T9	S.g Iron
Initial Weight	42.51	38.74	47.18	67.80
Final Weight	42.49	38.68	46.99	67.50
Weight Loss	0.02	0.06	0.19	0.30
Corrosion Rate Loss/Day	0.000948	0.00233	0.00916	0.0083

5.4.4 Corrosion Rate Conversion

The most used expression for corrosion rate in the US is the mpy (Miles per year).

To convert corrosion rate (corrosion rate conversion) between the mpy and the equivalent in the metric unit mm/y (millimeter per year).

1 mpy= 0.024 mm/y =22.4 microns/year

To calculate the corrosion rate from metal loss:

Mm/y=87.6*(W/DAT)

 Where:

W=weight loss in milligrams

D=metal density in g/cm^3

A=area of sample in cm^2

T=time of exposure of the metal sample in hours

CHAPTER 6
PARAMETER OPTIMIZATION USING FRICTION SURFACING

6.1 Taguchi Method

Taguchi methods have become increasingly popular in recent years. The overall objective of the method is to produce high quality product at low cost to the manufacturer. Optimization of process parameters can improve quality characteristics. Basically, classical process parameter design is complex and a large no. of experiments has to be carried out. Thus, Taguchi proposed an experimental design that uses orthogonal arrays to study the entire process parameters space with a small no. of experiments only. It helps to determine the factors that mostly affect the product quality. Taguchi recommends the use of loss function to measure the deviation of the quality characteristics from the desired value. Usually, there are three categories of quality charactcristics in the analysis of S/N ratio, i.e., the lower-the-better, higher-the-better, and the nominal-the-better. A large S/N ratio corresponds to better quality characteristic. Therefore the optimal level of the process parameters is the level with the highest S/N ratio.

6.2 Analysis of Variance (ANOVA)

The null hypothesis is that the treatment effects are all the same and the alternative is that they were not all the same (at least one of them differs significantly from the others). ANOVA helps to determine if any of the sources significantly affect the variability of the outcome being studied. In ANOVA, a dependent variable is measured under experimental conditions identified by independent variables. The variation in the response is due to effects in the classification, with random error accounting for remaining variation. The ANOVA procedure is one of several procedures available in statistical software for analysis of variance.

6.3 Response Surface Methodology

Response surface methods are used to examine the relationship between one or more response variables and a set of quantitative experimental variables or factors. These methods are often employed after identifying a "vital few controllable factors and to find the factor settings that optimize the response.

Response surface methods may be employed to

- Find factor settings (operating conditions) that produce the best response.
- Find factor settings that satisfy operating or process specifications.

- Identify new operating conditions that produce demonstrated improvement in product quality over the quality achieved by current conditions.

6.4 Statistical Software/Tool – MINITAB V.15

Minitab is a statistics package developed and distributed by Minitab Inc., Minitab works with data in worksheets of rows and columns. It has over 200 commands to perform various manipulations, statistical analyses, and graphics on data in the worksheet. This project uses Minitab Version 15 for the statistical analysis and optimization.

CHAPTER 7
OPTIMIZATION OF OUTPUT PARAMETERS

7.1 Purpose of ANOVA

The purpose of the analysis of variance (ANOVA) is to investigate which design parameters significantly affect the quality characteristic. This is to accomplished by separating the total variability of the S/N ratios, which is measured by the sum of the squared deviations from the total mean S/N ratio, into contributions by each of the design parameters and the error. First, the total sum of squared deviations SST from the total mean S/N ratio nm can be calculated as $SS_T = \sum (n_i - n_m)^2$

where n is the number of experiments in the orthogonal array and n_i is the mean S/N ratio for the ith experiment.

7.2 Optimization of Coating Width Parameters

7.2.1 Regression Analysis for Width

The regression equation is

D = 80.3 + 0.0123 A - 0.0374 B - 1.45 C

Table 7.1

Predictor	Coef	SE Coef	T	P
Constant	80.265	7.063	11.36	0.000
A	0.012275	0.003731	3.29	0.022
B	-0.037379	0.003725	-10.04	0.000
C	-1.4478	0.3725	-3.89	0.012

S = 0.6873   R-Sq = 96.1%   R-Sq(adj) = 93.8%

7.2.2 Analysis of Variance for Width

Table 7.2

Source	DF	SS	MS	F	P

Regression	3	14086.7	4695.6	41.11	0.001
Residual error	5	571.1	114.2		
Total	8	14657.8			

Estimated Modal Coefficients for S/N ratio (Width)

Table 7.3

Term	Coef	SE Coef	T	P
Constant	23.89	0.0318	662.21	0.000
A 500	-0.3020	0.05182	-5.598	0.027
A 550	-0.0325	0.05182	-0.611	0.571
B 1750	0.8621	0.05182	15.119	0.006
B 1800	0.1701	0.05182	3.182	0.069
C 2.0	0.4188	0.05182	8.078	0.019
C 1.0	-0.1312	0.05182	-2.498	0.120

S = 0.1136   R-Sq = 97.3%   R-Sq(adj) = 98.7%

Analysis of Variance for SN ratios (Width)

Table 7.4

Source	DF	Seq SS	Adj SS	Adj MS	F	P	%Contribution
A	2	0.61025	0.61025	0.30513	24.15	0.040	8.815
B	2	5.41976	5.41976	2.70988	214.49	0.005	78.294
C	2	0.86701	0.086701	0.43351	34.310	0.028	12.524
Residual error	2	0.02527	0.02527	0.01263			0.365
Total	8	6.92229					100

Estimated Modal Coefficients for Means (Width)

Table 7.5

Term	Coef	SE Coef	T	P
Constant	17.498	0.67323	211.491	0.000
A 500	-0.5371	0.10899	-4.756	0.045

A 550	-0.1191	0.10899	-1.011	0.427
B 1750	1.7099	0.10899	14.603	0.005
B 1800	0.2916	0.10899	2.373	0.130
C 2.0	0.8802	0.10899	7.290	0.020
C 1.0	-0.3119	0.10899	-2.649	0.114

S = 0.2504   R-Sq = 99.2%   R-Sq(adj) = 98.3%

Analysis of Variance for Means (Width)

Table 7.6

Source	DF	Seq SS	Adj SS	Adj MS	F	P	%Contribution
A	2	2.2840	2.2840	1.1420	18.26	0.032	8.396
B	2	21.2880	21.4880	10.6440	170.18	0.006	78.255
C	2	3.5060	3.5060	1.7530	28.03	0.034	12.888
Residual error	2	0.1251	0.1251	0.0625			0.459
Total	8	27.2032					100

7.2.3 Response Tables for S/N Ratio (Width)

Response table for Signal to Noise Ratios – Larger is better

Table 7.7

Level	Axial Load	Rotational Speed	Traverse Speed
1	24.55	25.70	25.28
2	24.81	25.02	24.71
3	25.18	23.82	24.55
Delta	0.64	1.88	0.72
Rank	3	1	2

Based on the above table we can observe that the rotational speed has contributed the most in affecting the width values and thus occupies the first position.

Response table for Means (Width)

Table 7.8

Level	Axial Load	Rotational Speed	Traverse Speed
1	17.01	19.28	18.43
2	17.44	17.85	17.25
3	18.27	15.55	17.00
Delta	1.22	3.73	1.43
Rank	3	1	2

Based on the above table we can observe that the rotational speed has contributed the most in affecting the width values and thus occupies the first position.

7.2.4 Main Effects Plot for Means

When performing a statistical analysis, one of the simplest graphical tools is a main effects plot. This plot shows the average outcome for each value of each variable, combining the effects of other variables.

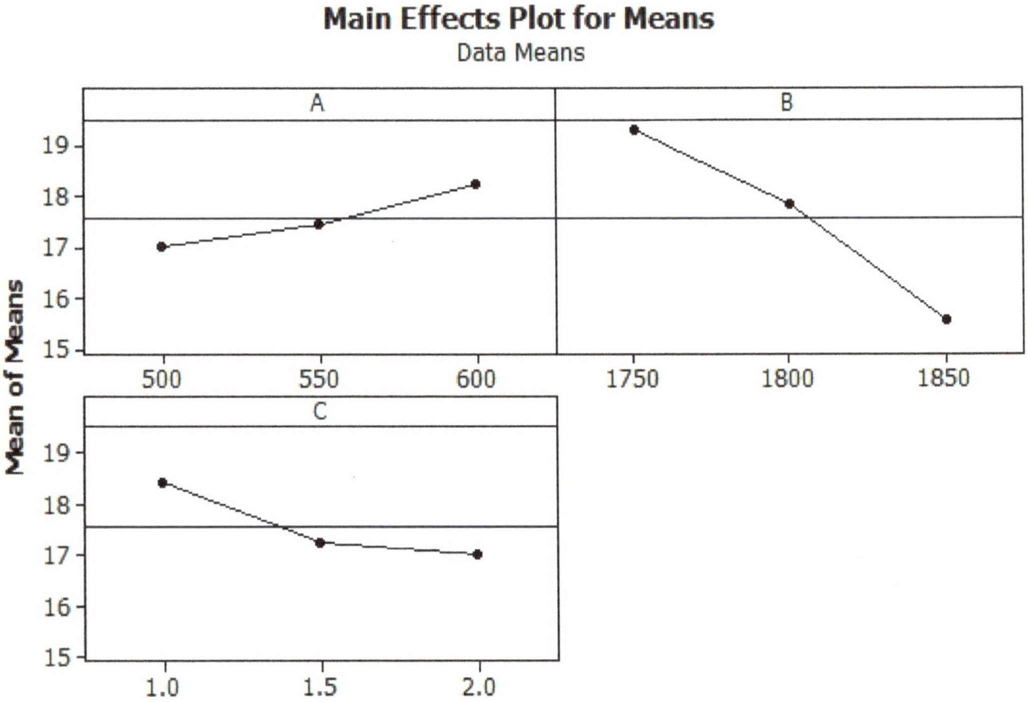

Fig 7.1

The above graph (figure 7.1) shows the main effects plot for means where the optimum parameter will be based on the highest peak at each parameter. From the figure, it can be seen that highest width is obtained for 600 kg axial load, 1750 rpm rotational speed and 1.0 mm/s traverse speed.

7.2.5 Main Effects Plot for Signal to Noise Ratio

Signal-to-noise ratio often abbreviated as SNR or S/N is a measure used in science and engineering to quantify how much a signal has been corrupted by noise. The SN ratio transforms several repetitions into one value which reflects the amount of variation present and the mean response.

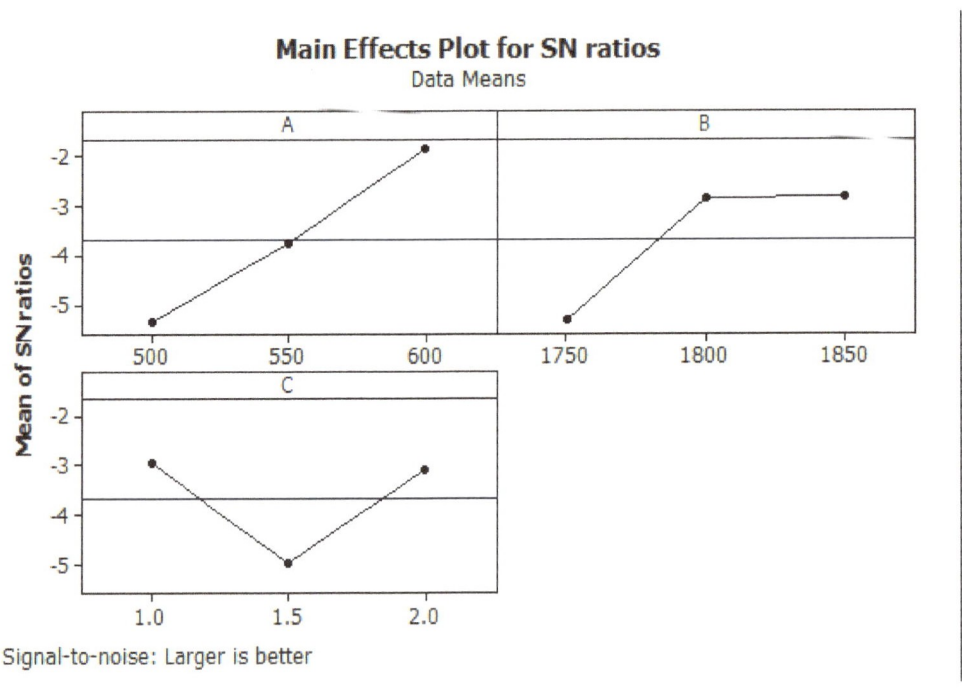

Fig 7.2

The above graph (figure 7.2) shows the main effects plot for the S/N ratio where the optimum parameter will be based on the highest peak at each parameter. From the figure it can be seen that width is obtained for 600 kg axial load, 1850 rpm rotational speed and 1.0 mm/s traverse speed.

7.2.6 Contour Plot

In a contour plot, the values for two variables are represented on the x and y axes, while the values for a third variable are represented by shaded regions, called contours. A contour plot is like a topographical map in which x, y and z values are plotted instead of longitude, latitude and altitude.

The graph (figure 7.3) shows the contour plot of width (in mm) vs. axial load (in N) and rotational speed (in rpm). It can be seen that maximum width can be obtained from minimum and maximum axial load and minimum rotational speed.

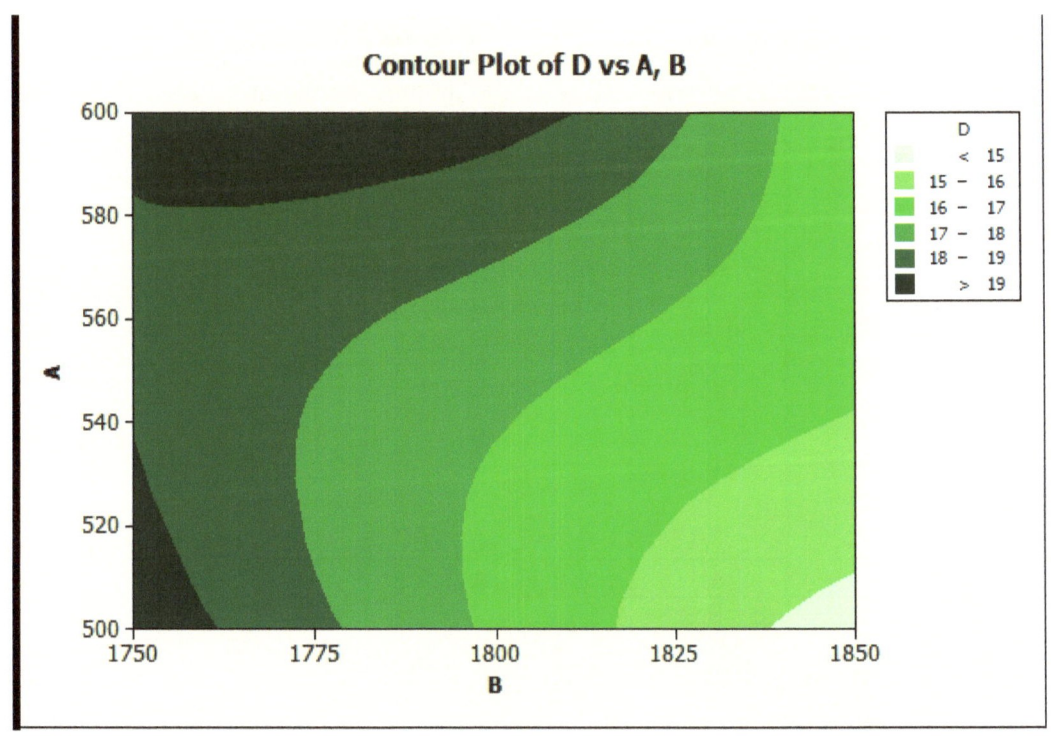

Fig 7.3

The graph (figure 7.4) shows the contour plot of width (in mm) vs. rotational speed (in rpm) and traverse speed (in mm/s). It can be seen that maximum width can be obtained until intermediate rotational speed and minimum and maximum traverse speed.

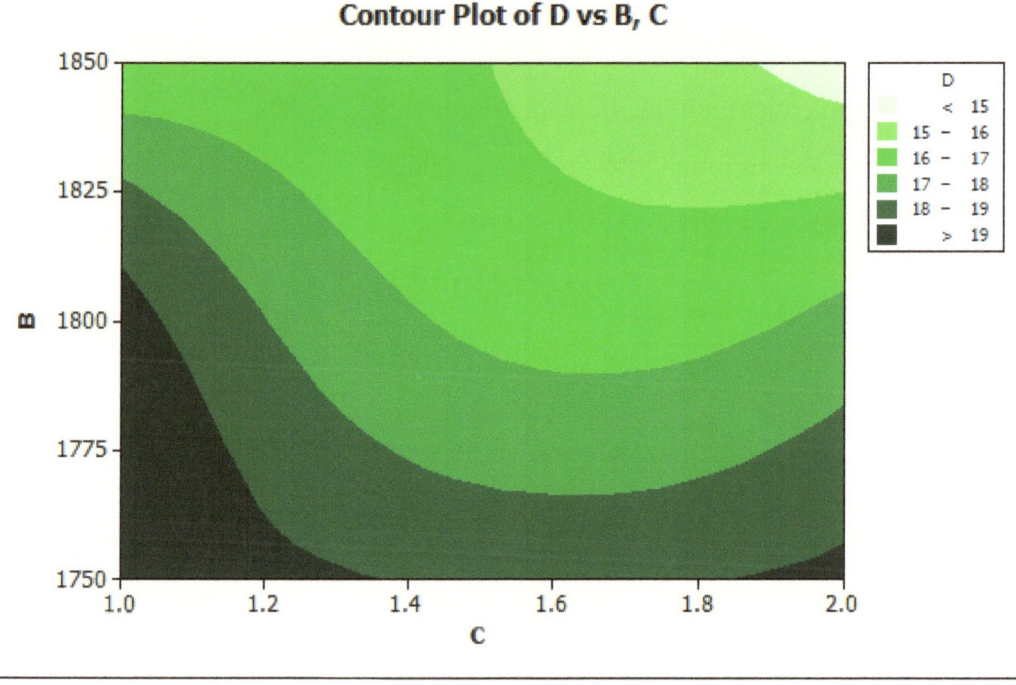

Fig 7.4

The graph (figure 7.5) shows the contour plot of width (in mm) vs. axial load (in N) and traverse speed (in mm/s). It can be seen from the figure that maximum width can be found in minimum and maximum axial load and traverse speed.

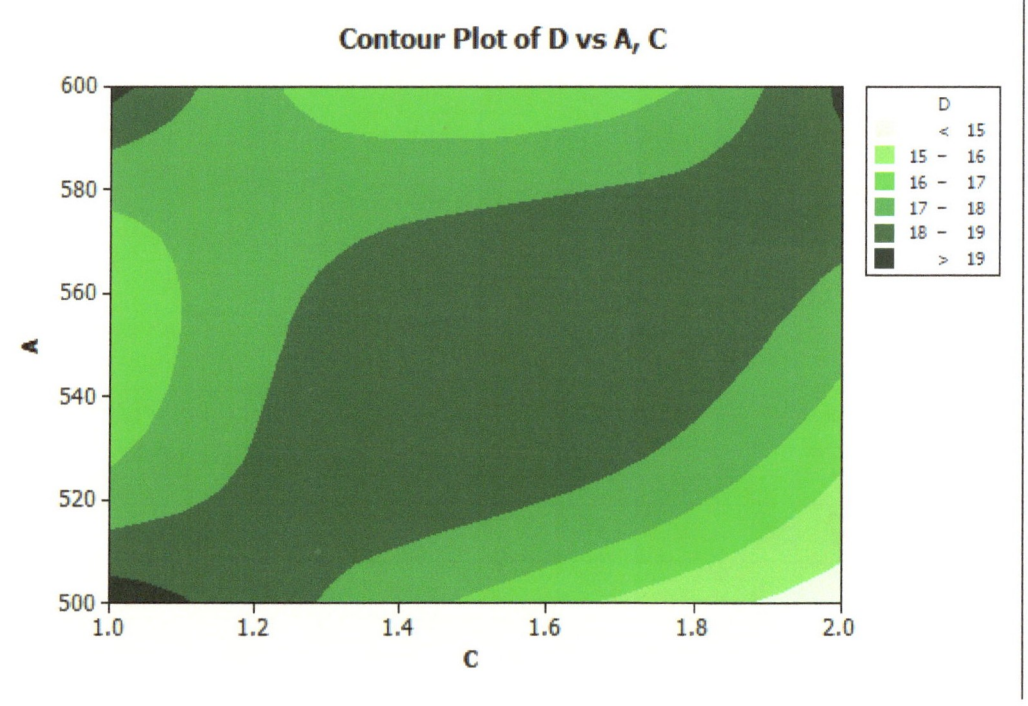

Fig 7.5

7.2.7 Residual Plot

Probability plot helps to determine whether a particular distribution fits your data or to compare different sample distributions. If the distribution fits the data:

- The plotted points will roughly form a straight line.
- The plotted points will fall close to the fitted distribution line.

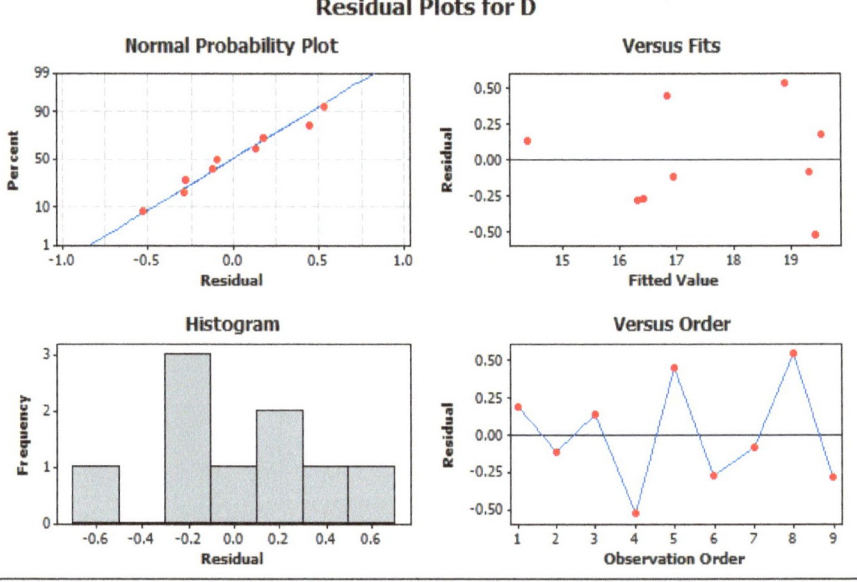

Fig 7.6

The (blue line) distribution line at the centre of the graph shows the desired width value in the normal probability plot. The trials (red dots) which lie close to the distribution line in the graph show the degree of closeness of the values.

7.2.8 Predicted Width

Table 7.9

Trials	Obtained Width	Predicted Width
T1	19.72	19.45
T2	16.83	16.02
T3	14.49	14.12
T4	18.9	19.3862
T5	17.29	16.8443

T6	16.14	16.4543
T7	19.23	19.3296
T8	19.44	17.989
T9	16.04	16.1124

From our experiment it is inferred that the obtained width and the predicted width are almost similar.

7.3 Optimization of Thickness

7.3.1 Regression Analysis for Thickness

The regression equation is

D = - 3.77 + 0.00277 A + 0.00163 B - 0.0267 C

Table 7.10

Predictor	Coef	SE Coef	T	P
Constant	-3.7743	0.4973	-7.59	0.001
A	0.0027731	0.0002627	10.56	0.000
B	0.0016313	0.0002622	6.22	0.002
C	-0.02668	0.02622	-1.02	0.356

S=0.752474   R-Sq = 96.8%   R-Sq(adj) =94.9%

7.3.2 Analysis of Variance for Thickness

Table 7.11

Source	DF	SS	MS	F	P
Regression	3	85.960	28.653	50.06	0.000
Residual error	5	2.831	0.566		
Total	8	88.791			

Estimated Modal Coefficients for S/N ratio (Thickness)

Table 7.12

Term	Coef	SE Coef	T	P
Constant	7.1487	0.2388	30.911	0.000
A 500	-0.5011	0.3219	-1.611	0.284

51

A 550	0.4988	0.3219	1.684	0.230
B 1750	1.3007	0.3219	4.077	0.059
B 1800	0.3667	0.3219	1.1209	0.401
C 2.0	1.1277	0.3219	2.990	0.067
C 1.0	0.1870	0.3219	0.603	0.619

S = 0.6345   R-Sq = 93.4%   R-Sq(adj) = 88.34%

Analysis of Variance for SN ratios (Thickness)

Table 7.13

Source	DF	Seq SS	Adj SS	Adj MS	F	P	%Contribution
A	2	1.6143	1.6143	0.8072	1.76	0.362	6.387
B	2	13.5229	13.5229	6.7615	14.74	0.064	53.504
C	2	9.2199	9.2199	4.6099	10.05	0.090	36.479
Residual error	2	0.9172	0.9172	0.4586			3.628
Total	8	25.2744					100

Estimated Modal Coefficients for Means (Thickness)

Table 7.14

Term	Coef	SE Coef	T	P
Constant	2.32599	0.02718	62.342	0.000
A 500	-0.08020	0.04215	-2.323	0.254
A 550	0.10813	0.04215	2.331	0.125
B 1750	0.33420	0.04215	4.324	0.032
B 1800	0.06754	0.04215	1.231	0.375
C 2.0	0.29488	0.04215	5.864	0.032
C 1.0	0.04201	0.04215	0.666	0.612

S = 0.1133   R-Sq = 92.4%   R-Sq(adj) = 94.5%

Analysis of Variance for Means (Thickness)

Table 7.15

Source	DF	Seq SS	Adj SS	Adj MS	F	P	%Contribution
A	2	0.06287	0.06287	0.03144	2.67	0.272	4.301
B	2	0.81891	0.81891	0.40945	34.79	0.028	56.022
C	2	0.55642	0.55642	0.27821	23.64	0.041	38.065
Residual error	2	0.02354	0.02354	0.01177			1.610
Total	8	1.46174					100

7.3.3 Response Tables for S/N Ratio (Thickness)

Response table for Signal to Noise Ratios – Larger is better

Table 7.16

Level	Axial Load	Rotational Speed	Traverse Speed
1	-5.620	-5.072	-3.811
2	-4.278	-3.929	-4.103
3	-1.974	-2.872	-3.959
Delta	3.646	2.201	0.292
Rank	1	2	3

Based on the above table we can observe that the rotational speed has contributed the most in affecting the thickness values and thus occupies the first position.

Response table for Means (Thickness)

Table 7.17

Level	Axial Load	Rotational Speed	Traverse Speed
1	0.5262	0.5657	0.6621
2	0.6150	0.6480	0.6442
3	0.8006	0.7282	0.6355
Delta	0.2744	0.1625	0.0266

Rank	1	2	3

Based on the above table we can observe that rotational speed has contributed the most in affecting the thickness values and thus occupies the first position.

7.3.4 Main Effects Plot for Means

When performing a statistical analysis, one of the simplest graphical tools is a main effects plot. This plot shows the average outcome for each value of each variable, combining the effects of other variables.

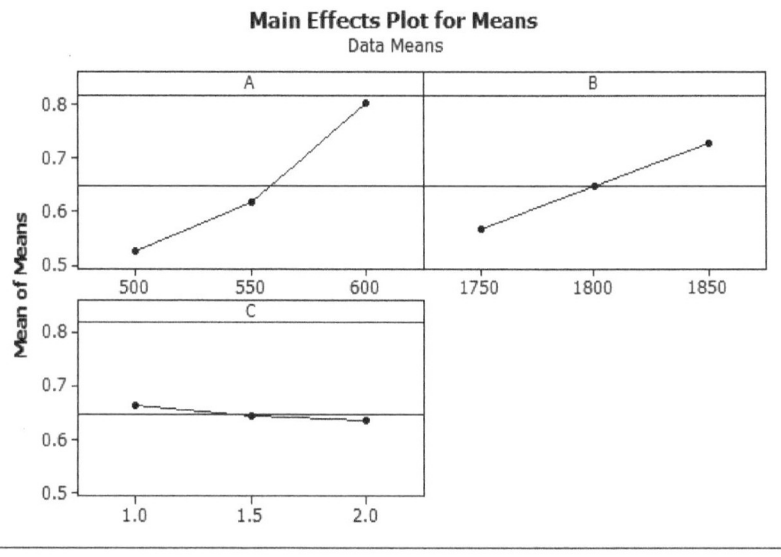

Fig 7.7

The above graph (figure 7.7) shows the main effects plot for means where the optimum parameter will be based on the highest peak at each parameter. From the figure, it can be seen that highest thickness is obtained for 600 kg axial load, 1850 rpm rotational speed and 1.0 mm/s traverse speed.

7.3.5 Main Effects Plot for Signal to Noise Ratio

Signal-to-noise ratio often abbreviated as SNR or S/N is a measure used in science and engineering to quantify how much a signal has been corrupted by noise. The SN ratio transforms several repetitions into one value which reflects the amount of variation present and the mean response.

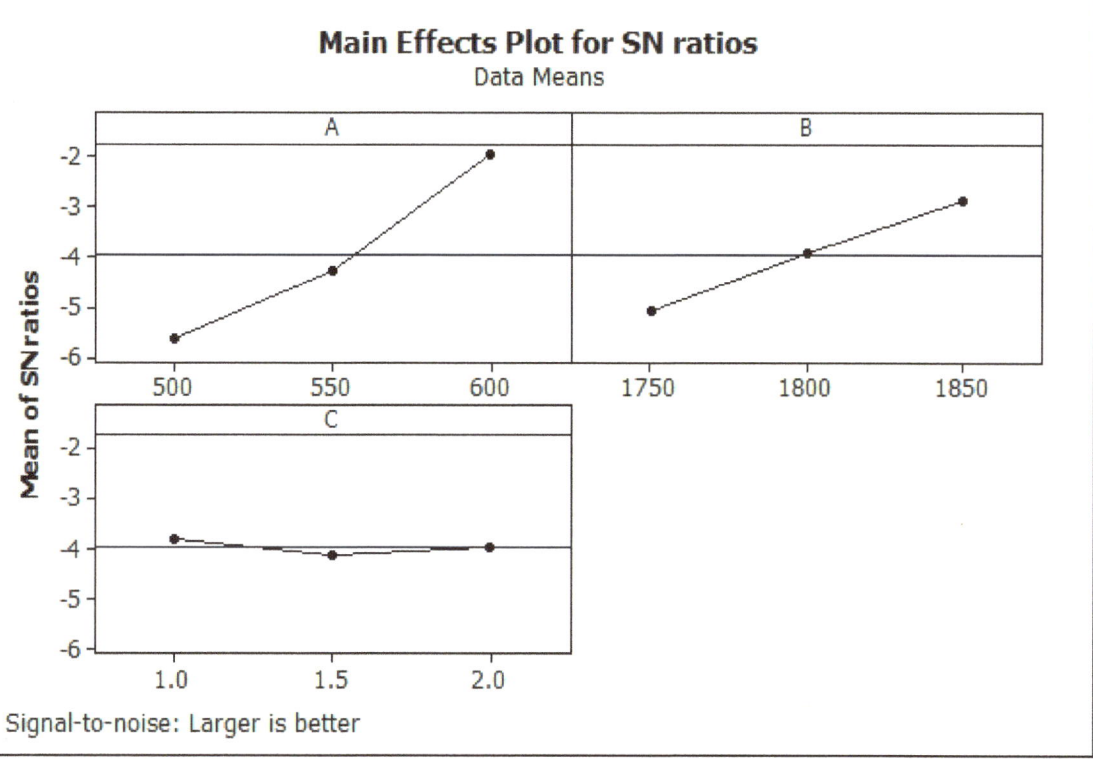

Fig 7.8

The above graph (figure 7.8) shows the main effects plot for the S/N ratio where the optimum parameter will be based on the highest peak at each parameter. From the figure it can be seen that highest thickness is obtained for 600 kg axial load, 1850 rpm rotational speed and 1.0 mm/s traverse speed.

7.3.6 Contour Plot

In a contour plot, the values for two variables are represented on the x and y axes, while the values for a third variable are represented by shaded regions, called contours. A contour plot is like a topographical map in which x, y and z values are plotted instead of longitude, latitude and altitude.

The graph (figure 7.9) shows the contour plot of thickness (in mm) vs. axial load (in N) and rotational speed (in rpm). It can be seen from the figure that maximum thickness can be obtained at the maximum axial load and minimum rotational speed.

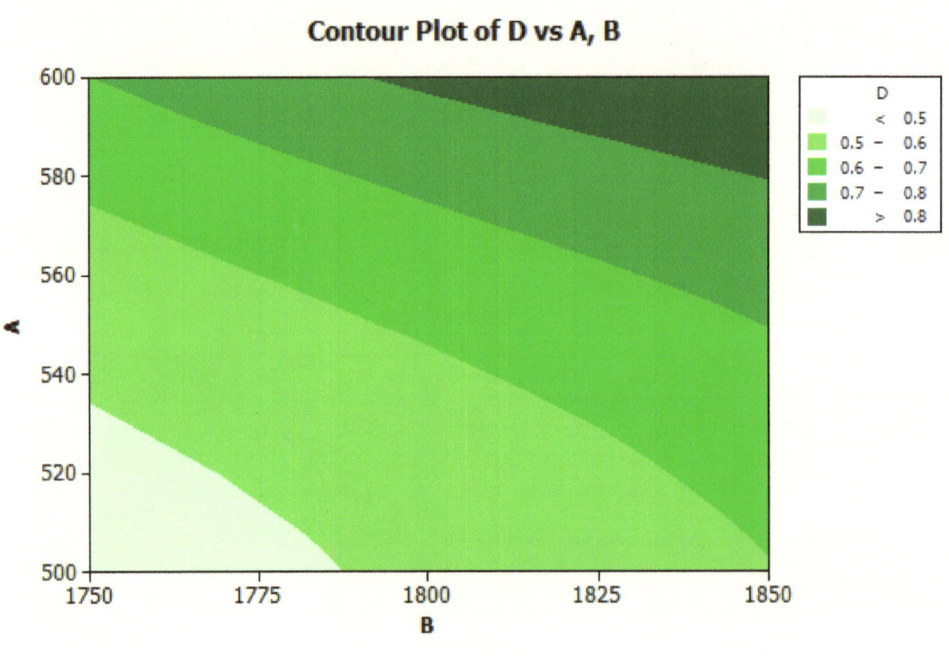

Fig 7.9

The graph (figure 7.10) shows the contour plot of thickness (in mm) vs rotational speed (in rpm) and traverse speed (in mm/s). It can be seen from the figure that the maximum thickness can be obtained at minimum rotational and traverse speed.

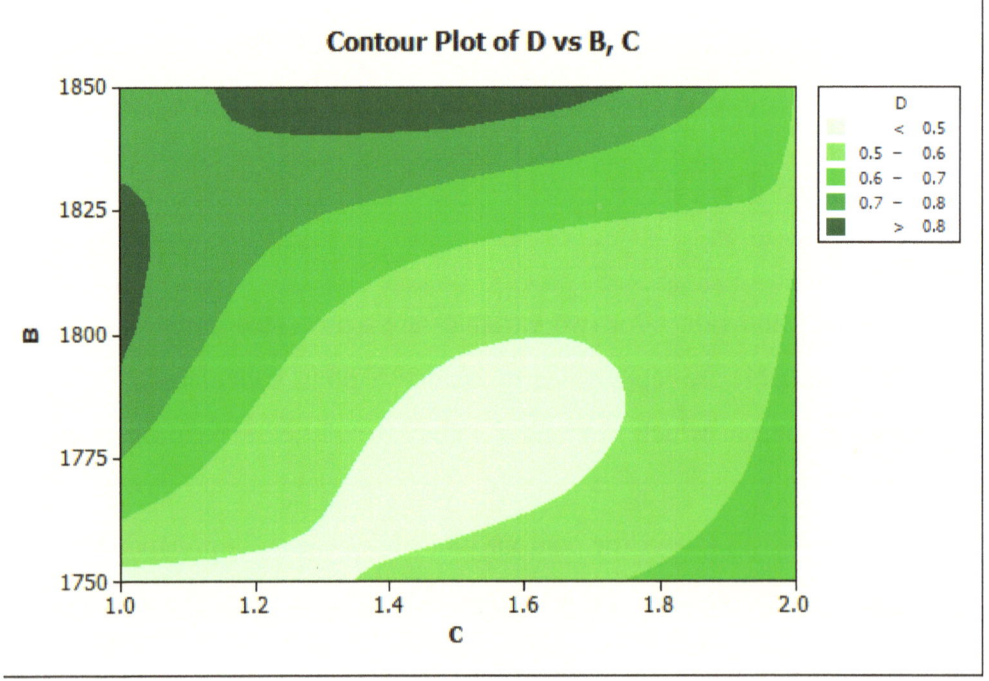

Fig 7.10

The graph (figure 7.11) shows the contour plot of thickness (in mm) vs. axial load (in N) and traverse speed (in mm/s). It can be seen that maximum thickness can be obtained at minimum traverse and axial load.

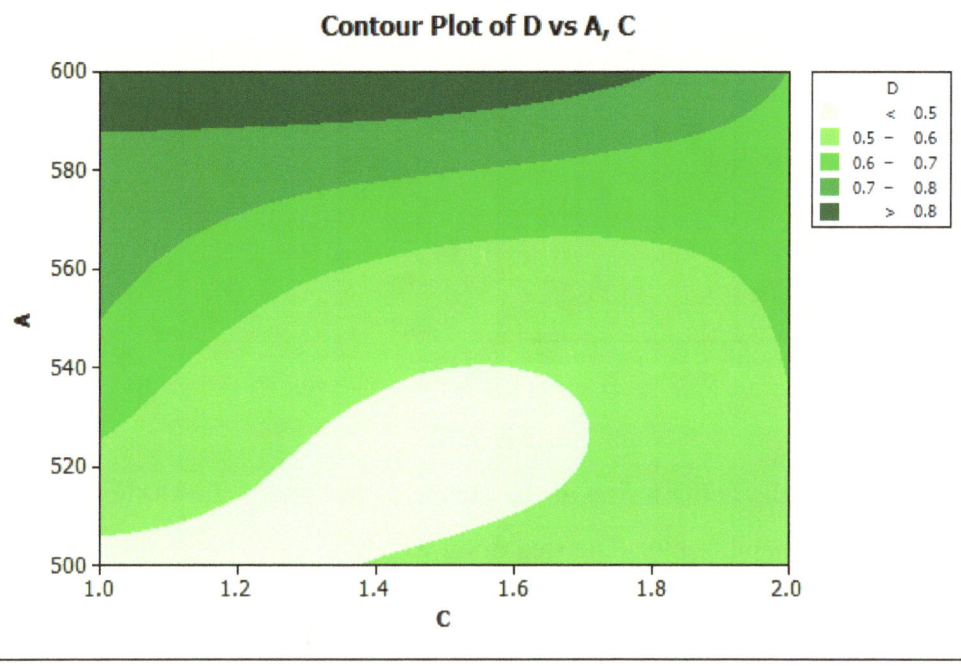

Fig 7.11

7.3.7 Residual Plot

Probability plot helps to determine whether a particular distribution fits your data or to compare different sample distributions. If the distribution fits the data:

- The plotted points will roughly form a straight line.
- The plotted points will fall close to the fitted distribution line.

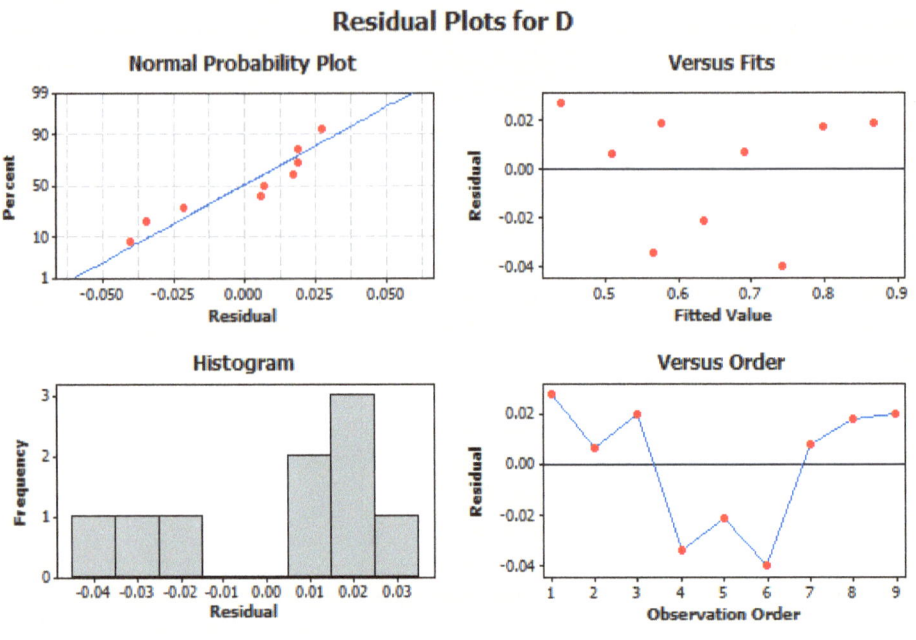

Fig 7.12

The (blue line) distribution line at the centre of the graph shows the desired thickness value in the normal probability plot. The (red dots) which lie close to the distribution line in the graph show the degree of closeness of the values.

7.3.8 Predicted Thickness

Table 7.18

Trials	Obtained Thickness	Predicted Thickness
T1	2.9	2.03
T2	2.36	2.1198
T3	1.45	1.5986
T4	2.725	2.4568
T5	2.21	2.5520
T6	2.355	2.53036
T7	2.325	2.02856
T8	2.55	2.5836
T9	1.95	1.2265

From the above table it is inferred that the thickness obtained from our experiment is almost similar to the predicted thickness.

CHAPTER 8
CONCLUSION

- Experimental results show that the friction surfacing could be used as a method for obtaining coatings of dissimilar materials. Friction surfacing is the best method for obtaining deposits of stainless steel over ductile iron for critical applications. Adequate bond strength and good coating integrity of deposit is obtained by optimizing of process parameters.

- The width of the deposit is always less than the diameter of the mechatrode used and lies between 0.6 to 0.95of it and its value depends on process parameters used. This response is more important in depositing the consumable in slots.

- The microstructure reveals good bond between stainless steel and ductile iron which is obtained by the results of the combined forging and shear action of mechatrode at the plastic state with ductile iron. The interface layer zone is the intermixed materials of substrate and mechatrode.

- The deposit observed by the microscope showed dense, clear and fine microstructure of ferrite and pearlite on ductile iron side which clearly proves the superiority of the process.

- There were no cracks observed in the HAZ, showing the suitability of the parameters selected to give controlled heat input. Integrity of the deposit is excellent with good metallurgical bond.

- Corrosion test and bend tests results proved that this method is can be for manufacture of petrochemical vessels, pumps for chemicals and other corrosion resistant applications. There is tremendous scope to extend this process to other dissimilar metal combinations for protection against wear and corrosion.

- The width from the trial 1 of our experiment is better comparing it with the other trials considering larger is better.

- The thickness of trial 4 of our experiment is better comparing it with the other trials considering larger is better.

- The result obtained from trial 1 is better in comparison with the remaining trials. This is obtained by taking both experimental width and thickness into consideration.

CHAPTER 9
SCOPE OF FUTURE WORK

- Study the characteristics and mechanical properties of the friction surfaced stainless steel deposits, when performed in water and inert gas.

- Analysis of the stainless steel deposit over the ductile iron when it is made in the form of pad. Determining the process parameters when the deposit is made over the pad with adjacent layer and multilayer.

- Joining of materials by friction surfacing which are having large differences in thermal expansion. When these materials joined, high stresses are developed during the cooling. Hence intermediate expansion material is expansion material is required to allow for the transition from high to low thermal expansion materials. Example refractory of metals, ceramics, and low-expansion iron-nickel and iron nickel- cobalt alloys may fail or be highly stressed during cooling when welded to high -expansion material such as austenitic stainless steels and nickel-base and cobalt-base super alloys.

REFERENCES

References:

[1] G. Madhusudhan Reddy and T. Mohandas, "Friction Surfacing of Metallic Coatings on Steels", Proceedings of the International Institute of Welding International Congress 2008, Chennai, India, January 2008, pp 1197 – 1213.

[2] I. Voutchkov, B. Jaworski, V. I. Vitanov, and G. M. Bedford, "An integrated approach to friction surfacing process optimization", Surface and Coatings Technology, Vol. 141, 2001, pp 26-33.

[3] X. M. Liu, Z. D. Zou, Y. H. Zhang, S. Y. Qu and X. H. Wang, "Transferring mechanism of the coating rod in friction surfacing", Surface and Coating Technology, Vol.202, 2008, pp 1889-1894.

[4] Margam Chandrasekaran, Andrew William Batchelor and Sukumar Jana, "Friction surfacing of metal coatings on steel and aluminium substrate", Journal of Materials Processing Technology, Volume 72, 15 December 1997, Pp 446-452.

[5] K. Prasad Rao, A. Veera Sreenu, H. Khalid Rafi, M.N. Libin and Krishnan Balasubramaniam, " Tool steel and copper coatings by friction surfacing – A thermography study", Journal of Materials Processing Technology, Volume 212, Issue 2, February 2012, Pp 402-407.

[6] Ramesh Puli and G.D. Janaki Ram, "Wear and corrosion performance of AISI 410 martensitic stainless steel coatings produced using friction surfacing and manual metal arc welding", Surface and Coatings Technology, Volume 209, 25 September 2012, Pp 1-7.

[7] H. Khalid Rafi, G.D. Janaki Ram, G. Phanikumar and K. Prasad Rao, "Microstructural evolution during friction surfacing of tool steel H1", Materials & Design, Volume 32, Issue 1, January 2011, Pp 82-87.

[8] H. Khalid Rafi, G.D. Janaki Ram, G. Phanikumar and K. Prasad Rao, "Friction surfaced tool steel (H13) coatings on low carbon steel: A study on the effects of process parameters on coating characteristics and integrity", Surface and Coatings Technology, Volume 205, Issue 1, 25 September 2010, Pp 232–242.

[9] H.Khalid Rafi, G.D.Janaki Ram, G.Phanikumar and K.Prasad Rao, " Friction Surfacing of Austenitic Stainless Steel on Low Carbon Steel:Studies on the Effects of Traverse Speed", Proceedings of the World Congress on Engineering, 2010.

[10] M.Chandrasekaran, A.W.Batchelor and S.Jana, "Study of the interfacial phenomena during friction surfacing of aluminium with steels", Journal of materials scienceVolume:32,Date:1997, Pp 6055-6062.

[11] H. Khalid rafi, N.Kishore babu, G.Phanikumar and K.Prasad Rao "Microstructural Evolution During Friction Surfacing of Austenitic Stainless Steel AISI 304 on Low Carbon Steel",The Minerals, Metals & Materials Society and ASM International 2012Volume:44A Date:2013,Pp 345-350.

[12] Ramesh Puli, E. Nandha Kumar G.D and Janaki Ram , "Characterization of friction surfaced martensitic stainless steel(AISI 410) coatings",Transactions of The Indian Institute of Metals Volume:64,Date:2011, Pp 41-45.

[13] M.Chandrasekaran, A.W.Batchelor and S.Jana "Study of the interfacial phenomena during friction surfacing of mild steel with tool steel and inconel", Journal of materials science, Volume:32,Date:1997, Pp 6055-6062.

[14] J. Gandra, R.M. Miranda and P. Vilaça, "Performance analysis of friction surfacing", Journal of Materials Processing Technology, Volume 212, Issue 8, August 2012, Pp 1676-1686.

[15] G.M. Bedford, V.I. Vitanov and I.I. Voutchkov , "On the thermo-mechanical events during friction surfacing of high speed steels", Surface and Coatings Technology, Volume 141, Issue 1, 4 June 2001, Pp 34-39.

[16] V.I Vitanov, I.I Voutchkov, G.M Bedford, " Neurofuzzy approach to process par ameter selection for friction surfacing applications", Surface and Coatings Technology, Volume 140, 1 June 2001, Pp 256-262.

[17] D.Govardhan, A.C.S. Kumar, K.G.K. Murti and G. Madhusudhan Reddy, "characterization of austenitic stainless steel friction surfaced deposit over low carbon steel", Materials & Design, Volume 36, April 2012, Pp 206-214.

[18] Rick Greenough "Friction surfacing for Multi Sectorial applications",Frictec-Applicataion development –friction surfacing technology, Nov 2006.

[19] Hoshihiro Yamashita and Kazuhiro Fujita "Newly Developed Repairs on Welded Area of LWR Stainless steel by friction surfacing ", Journal of nuclear science and technology, Vol.38, No10, Pp 896-900.

www.ingramcontent.com/pod-product-compliance
Lightning Source LLC
Chambersburg PA
CBHW040742200526
45159CB00023B/1461